Analysis of the Air Force Logistics Enterprise

Evaluation of Global Repair Network Options for Supporting the F-16 and KC-135

T0195515

Ronald G. McGarvey, Manuel Carrillo, Douglas C. Cato, Jr., John G. Drew, Thomas Lang, Kristin F. Lynch, Amy L. Maletic, Hugh G. Massey, James M. Masters, Raymond A. Pyles, Ricardo Sanchez, Jerry M. Sollinger, Brent Thomas, Robert S. Tripp, Ben D. Van Roo

Prepared for the United States Air Force

Approved for public release; distribution unlimited

PROJECT AIR FORCE

The research described in this report was sponsored by the United States Air Force under Contract FA7014-06-C-0001. Further information may be obtained from the Strategic Planning Division, Directorate of Plans, Hq USAF.

Library of Congress Cataloging-in-Publication Data

Analysis of the Air Force logistics enterprise : evaluation of global repair network
    options for supporting the F-16 and KC-135 / Ronald G. McGarvey ... [et al.].
        p. cm.
    Includes bibliographical references.
    ISBN 978-0-8330-4740-3 (pbk.)
    1. Airplanes, Military—United States—Maintenance and repair—
Management—Evaluation. 2. United States. Air Force—Equipment—Maintenance
and repair—Management—Evaluation. 3. F-16 (Jet fighter plane)—Maintenance
and repair—Management—Evaluation. 4. F-16 (Jet fighter plane)—Maintenance
and repair—Costs—Evaluation. 5. KC-135 (Tanker aircraft)—Maintenance and
repair—Management—Evaluation. 6. KC-135 (Tanker aircraft)—Maintenance and
repair—Costs—Evaluation. 7. Logistics. I. McGarvey, Ronald G.

UG1243.A48 2009
358.4'1411—dc22

                                                                    2009031287

Published 2009 by the RAND Corporation
1776 Main Street, P.O. Box 2138, Santa Monica, CA 90407-2138
1200 South Hayes Street, Arlington, VA 22202-5050
4570 Fifth Avenue, Suite 600, Pittsburgh, PA 15213-2665
RAND URL: http://www.rand.org/
To order RAND documents or to obtain additional information, contact
Distribution Services: Telephone: (310) 451-7002;
Fax: (310) 451-6915; Email: order@rand.org

Preface

The Air Force has developed an extensive set of logistics resources and concepts to support training, deployment, employment, and redeployment of air, space, and cyber forces. Since the establishment of the Iraqi no-fly zones following the Persian Gulf War in 1991, the Air Force has been continuously engaged in rotational deployments, which have presented a very different set of challenges from those that were used to develop the current support posture, one still largely based on assumptions developed during the Cold War. Changes to the operational environment, such as rotational deployments of less-than-squadron-size operationally tasked units to unanticipated locations for unknown durations, may warrant changes to the Air Force's logistics infrastructure and concepts of operation, but the questions of "what changes" and "to what extent" remain unanswered.

Recognizing the importance of the logistics enterprise, the Air Force has initiated a set of interrelated transformation activities, including Air Force Smart Operations 21, a way of institutionalizing continuous process improvement; Expeditionary Logistics for the 21st Century (U.S. Air Force, 2005), an effort to transform current logistics processes to improve support to the operational units; and Repair Enterprise 21 (RE21),[1] which establishes an enterprisewide repair capability via the use of centralized intermediate repair facilities (CIRFs) for a subset of aircraft avionics and engines. However, ensuring that these efforts (which may at times promote conflicting objectives) remain integrated into a larger vision for transformation is a cause for concern.

---

[1] RE21 has since been subsumed by the Repair Network Integration initiative.

In light of these changes in the Air Force's operational environment, along with the potential for future reductions in the resources made available to the logistics enterprise, senior Air Force leaders have requested that RAND Project AIR FORCE undertake a comprehensive strategic reassessment of the entire Air Force logistics enterprise—to reidentify and rethink the basic issues and premises on which the Air Force plans, organizes, and operates its logistics enterprise, from a Total Force (TF) perspective—including the active-duty Air Force, along with the Air Force Reserve Command (AFRC) and Air National Guard (ANG). At a fundamental level, the logistics enterprise strategy must answer the following three questions:

- What will the logistics workload be?
- How should the logistics workload be accomplished?
- How should these questions be revisited over time?

To answer these questions, we have organized the Logistics Enterprise Analysis project around the following four tasks:

- a review of Office of the Secretary of Defense (OSD) programming guidance to determine projected logistics system workloads
- a structured analysis of scheduled and unscheduled maintenance tasks to strategically rebalance workloads among operating units and supporting network nodes
- a strategic reevaluation of the objectives and proper roles of contract and organic support in the logistics enterprise
- a top-down review of the management of logistics transformation initiatives to ensure that they are aligned with broader logistics objectives.

This monograph documents the results of analyses that address the first two tasks for the F-16 and KC-135 fleets. This work was conducted between August 2007 and July 2008. Subsequent publications will address the other tasks and examine other weapon systems.

The monograph shows how operational units can be reconfigured to support launch and recovery operations, with "heavy maintenance," such as phase inspections for fighter aircraft, being provided by an enterprise network of centralized repair facilities (CRFs). Components would also be supplied to operational units from CRFs that specialize in repairing these assets, thus removing most backshop resources from operational units. Our analyses address the costs and benefits of an enterprise approach configured to support the TF. After we presented our initial TF analysis results, we were asked to evaluate an enterprise option that would be used to support only active-duty and AFRC forces. This monograph also contains that analysis.

The Deputy Chief of Staff for Logistics, Installations and Mission Support, along with the Vice Commander, Air Force Materiel Command (AFMC), sponsored this research, which was carried out in the Resource Management Program of RAND Project AIR FORCE under three projects: Enterprise Transformation Management for AFMC Umbrella Project, Global Materiel Management Strategy for the 21st Century Air Force, and Managing Workload Allocations in the USAF Global Repair Enterprise.

This monograph should interest logistics and operational personnel throughout the Department of Defense (DoD) and those involved in logistics manpower requirements determination.

Related RAND Corporation research includes the following:

- *Supporting Expeditionary Aerospace Forces: An Integrated Strategic Agile Combat Support Planning Framework*, Robert S. Tripp, Lionel A. Galway, Paul Killingsworth, Eric Peltz, Timothy Ramey, and John G. Drew (MR-1056-AF). This report describes an integrated combat support planning framework that may be used to evaluate support options on a continuing basis, particularly as technology, force structure, and threats change.
- *Supporting Expeditionary Aerospace Forces: An Analysis of F-15 Avionics Options*, Eric Peltz, Hyman L. Shulman, Robert S. Tripp, Timothy Ramey, and John G. Drew (MR-1174-AF). This report examines alternatives for meeting F-15 avionics maintenance requirements across a range of likely scenarios. The authors evaluate investments for new F-15 avionics intermediate shop test equipment against several support options, including deploying

maintenance capabilities with units, performing maintenance at forward support locations, and performing all maintenance at the home station for deploying units.

- *Supporting Expeditionary Aerospace Forces: Expanded Analysis of LANTIRN Options*, Amatzia Feinberg, Hyman L. Shulman, Louis W. Miller, and Robert S. Tripp (MR-1225-AF). This report examines alternatives for meeting Low-Altitude Navigation and Targeting Infrared for Night (LANTIRN) support requirements for Air and Space Expeditionary Force (AEF) operations. The authors evaluate investments for new LANTIRN test equipment against several support options, including deploying maintenance capabilities with units, performing maintenance at forward support locations, and performing all maintenance at continental United States (CONUS) support hubs for deploying units.

- *Supporting Expeditionary Aerospace Forces: Alternatives for Jet Engine Intermediate Maintenance*, Mahyar A. Amouzegar, Lionel A. Galway, and Amanda B. Geller (MR-1431-AF). This report evaluates the manner in which jet engine intermediate maintenance (JEIM) shops can best be configured to facilitate overseas deployments. The authors examine a number of JEIM support options, which are distinguished primarily by the degree to which JEIM support is centralized or decentralized. See also *Engine Maintenance Systems Evaluation (En Masse): A User's Guide*, Mahyar A. Amouzegar and Lionel A. Galway (MR-1614-AF).

- *Supporting Air and Space Expeditionary Forces: Analysis of Maintenance Forward Support Location Operations*, Amanda Geller, David George, Robert S. Tripp, Mahyar A. Amouzegar, and Charles Robert Roll, Jr. (MG-151-AF). This monograph discusses the conceptual development and recent implementation of maintenance forward support locations (also known as CIRFs) for the U.S. Air Force. The analysis focuses on the years leading up to and including the AF/IL CIRF test, which tested the operations of CIRFs in the European theater from September 2001 to February 2002.

- *Supporting Air and Space Expeditionary Forces: Lessons from Operation Iraqi Freedom*, Kristin F. Lynch, John G. Drew, Robert S.

Tripp, and Charles Robert Roll, Jr. (MG-193-AF). This monograph describes the expeditionary agile combat support experiences during the war in Iraq and compares those experiences with those associated with Joint Task Force Nobel Anvil in Serbia and Operation Enduring Freedom in Afghanistan. The monograph analyzes how combat support performed and how ACS concepts were implemented in Iraq, compares current experiences to determine similarities and unique practices, and indicates how well the ACS framework performed during these contingency operations.

- *Strategic Analysis of Air National Guard Combat Support and Reachback Functions*, Robert S. Tripp, Kristin F. Lynch, Ronald G. McGarvey, Don Snyder, Raymond A. Pyles, William A. Williams, and Charles Robert Roll, Jr. (MG-375-AF). This monograph analyzes transformational options for better meeting combat support mission needs for the AEF. The role the ANG may play in these transformational options is evaluated in terms of providing effective and efficient approaches to achieve the desired operational effects. Four Air Force mission areas are evaluated: CONUS CIRFs, civil-engineering deployment and sustainment capabilities, GUARDIAN[2] capabilities, and Air and Space Operations Center reachback missions.

- *Supporting Air and Space Expeditionary Forces: An Expanded Operational Architecture for Combat Support Planning and Execution Control*, Patrick Mills, Ken Evers, Donna Kinlin, and Robert S. Tripp (MG-316-AF). This monograph expands and provides more detail on several organizational nodes in earlier RAND work that outlined concepts for an operational architecture for guiding the development of Air Force combat support execution planning and control needed to enable rapid deployment and employment of AEFs (*Supporting Expeditionary Aerospace Forces: An Operational Architecture for Combat Support Execution Planning and Control*, James A. Leftwich, Robert S. Tripp, Amanda B. Geller, Patrick

---

[2] GUARDIAN (the Air National Guard Information Analysis Network) is an ANG system used to track and control execution of plans and operations, providing information such as funding and performance data.

Mills, Tom LaTourrette, Charles Robert Roll, Jr., Cauley Von Hoffman, and David Johansen [MR-1536-AF]). These combat support execution planning and control processes are sometimes referred to as combat support command and control processes.

- *Supporting Air and Space Expeditionary Forces: Analysis of CONUS Centralized Intermediate Repair Facilities,* Ronald G. McGarvey, James M. Masters, Louis Luangkesorn, Stephen Sheehy, John G. Drew, Robert Kerchner, Ben D. Van Roo, and Charles Robert Roll, Jr. (MG-418-AF). This monograph evaluates alternatives for establishing CIRFs for a set of aircraft avionics and engine components. The authors demonstrate that, for many components, use of CIRFs can improve system performance (via reductions in awaiting-maintenance queues) and reduce total system costs (via reductions in maintenance manpower requirements), because these effects outweigh the associated transportation pipelines and costs.

RAND Project AIR FORCE

RAND Project AIR FORCE (PAF), a division of the RAND Corporation, is the U.S. Air Force's federally funded research and development center for studies and analyses. PAF provides the Air Force with independent analyses of policy alternatives affecting the development, employment, combat readiness, and support of current and future aerospace forces. Research is conducted in four programs: Force Modernization and Employment; Manpower, Personnel, and Training; Resource Management; and Strategy and Doctrine. Integrative research projects and work on modeling and simulation are conducted on a PAF-wide basis.

Additional information about PAF is available on our Web site: http://www.rand.org/paf/

Contents

Figures

Tables

Summary

Background and Purpose

The Air Force has implemented a number of transformational initiatives since the advent of the AEF concept in 1998. Many of these initiatives have focused on incremental changes to the Air Force's logistics infrastructure and concepts of operation. In 2007, senior Air Force logisticians asked RAND to undertake a strategic reassessment of the Air Force's logistics enterprise to identify, using projections for the future operating environment, alternatives for appropriately rebalancing logistics resources and capabilities between operating units and support network nodes across the TF, including not only active duty (AD) forces but also the AFRC and ANG.

To meet this broad request, the Logistics Enterprise Analysis (LEA) project has been organized around the following four tasks:

- a review of OSD programming guidance to determine projected logistics system workloads
- a structured analysis of scheduled and unscheduled maintenance tasks to strategically rebalance workloads among operating units and supporting network nodes
- a strategic reevaluation of the objectives and proper roles of contract and organic support in the logistics enterprise
- a top-down review of the management of logistics transformation initiatives to ensure that they are aligned with broader logistics objectives.

The research presented in this monograph addresses only a part of the broader LEA project, namely, the first two tasks applied to the F-16 and KC-135 fleets. Subsequent publications will address the last two tasks and present enterprise rebalancing analyses for other weapon systems.

The strategic decisions made in these areas should recognize that key management options and important resource trade-offs occur in the following areas:

- **"Stockage" solutions versus "response" solutions.** Support of deployed and engaged forces has historically involved a blend of stocks (readiness spares packages, war readiness engines, prepositioned war reserve materiel, etc.) and responsive support (resupply, deployed intermediate-level maintenance, CIRF support, etc.). Strategic issues here involve our ability to forecast usage rates and requirements, in-theater footprint issues, and unit "ownership" versus "sharing" of assets.
- **Local maintenance versus network maintenance.** Flexibility exists in the location of maintenance activities that are not directly tied to sortie generation and recovery. For example, off-equipment component repair can be conducted on base in backshops[3] or off base at AFMC depot facilities, at CRFs, or at contractor facilities. This geographic dimension has important strategic considerations, including the extent of reliance on transportation, the speed of deployment, and maintenance manpower requirements, with differing risks associated with "self-sufficient" and "network" maintenance postures.

This study identifies alternatives that reallocate workload and resources between maintenance that is provided at the aircraft's operating location and maintenance that is provided from a flexible and robust network of CRFs. These alternatives provide equal or greater capability than the current system, with equal or fewer resources. Thus, the Air Force could use any savings either to increase its operational

---

[3] *Backshops* refers to one of three levels of Air Force aircraft maintenance. Historically, backshops perform intermediate-level maintenance; depots perform the highest level; and organizational, or "flightline," the lowest.

capability at no additional cost or to provide the same capability at less cost, capturing the savings associated with these economies to support other, more stressed areas than aircraft maintenance. An important aspect of this research is a commitment to identifying trade-offs among alternative solutions rather than advocating any single "best" option. Our goal is to inform Air Force leaders of the capability implications associated with varying levels of resource investment.

This monograph details the analysis that we performed to identify alternatives for rebalancing aircraft maintenance capabilities between unit-specific and network sources of repair for the F-16 and KC-135. The focus is on wing-level maintenance tasks, including sortie launch and recovery workloads, aircraft inspections, on-equipment maintenance to support removals and replacement of aircraft components, shop repair of replaceable components, and time-change technical orders. The analysis examined network-based alternatives, wherein each operational unit retains maintenance capabilities for performing aircraft launch and recovery and removal and replacement of failed components, with an enterprise network of CRFs providing major aircraft inspections (such as F-16 phase inspections) and component repair. We evaluated repair network options for supporting the TF and also alternatives for which the repair network supports only the AD and AFRC forces. The key trade-offs in this analysis occur between potential manpower economies of scale that can be realized via consolidation of workloads into a smaller number of sites and the transportation and facility costs associated with moving maintenance tasks away from the aircraft's operating location. We chose the F-16 and the KC-135 because of their dissimilarities in both logistics support requirements (e.g., F-16s have a relatively short phase-inspection interval and a limited flying range, while KC-135s have a relatively long inspection interval and a much longer flying range) and projected demands in support of future deployment scenarios (e.g., humanitarian relief operations require few, if any, F-16s but often have considerable demand for KC-135s).

We used a number of analytic tools to identify the resource requirements associated with these alternative maintenance constructs. The manpower requirements for unit-based "mission generation" (MG) maintenance were developed using a variety of sources, including

Unit Manning Document (UMD) and Unit Type Code (UTC) data describing the MG manpower as currently configured, along with extensive new simulation results obtained from the Logistics Composite Model (LCOM) to determine additional manpower requirements necessary to support our proposed maintenance restructuring. LCOM simulation results were also used to determine the CRF manpower requirements. We identified alternative repair network designs, consisting of the number, location, and size of CRFs, using an optimization model that minimizes the sum of the CRF manpower, transportation, and facility construction costs, subject to a variety of constraints. This optimization model considers the full range of CRF network alternatives, from fully decentralized solutions that retain CRF maintenance capabilities at all sites to fully centralized alternatives that consolidate all CRF capabilities at one site, and identifies the alternative that minimizes the total cost.

Results

F-16

Our analysis (see pp. 15–63) presents a method for optimizing resource allocations to provide a range of maintenance capabilities that either match or exceed those provided by the current structure. Suppose that the desired capability was the support of (1) a steady-state deployment of 10 percent of the combat-coded (CC) F-16 fleet into two theaters for an indefinite duration and (2) a surge deployment of 80 percent of the same fleet into two theaters. For this case, we identify an alternative, presented in Table S.1, that enhances the capability of F-16 maintenance units by transferring 1,900 manpower positions out of backshop maintenance,[4] made possible by centralizing certain backshop workloads, and moving these positions into MG maintenance, giving each CC squadron the ability to conduct split operations, in which F-16 squadrons have

---

[4] In Table S.1, these positions are in the shaded cell in the "TF Repair Network" column.

Table S.1
Option 1: F-16 Increased Operational Effectiveness

Operation	Manpower Authorization		
	Current System	AD/AFRC- Only Repair Network	TF Repair Network
Group and MOS: FY 2008 UMD	3,363	3,363	3,363
AMXS			
FY 2008 UMD	11,143	11,143	11,143
Moved from CMS and EMS		1,046	1,884
Split operations plus-up		844	1,896
CMS and EMS			
Propulsion and avionics: FY 2008 UMD	2,863	2,863	2,863
Age and munitions: FY 2008 UMD	4,093	4,093	4,093
Phase and related: FY 2008 UMD	6,221	2,714	
CRF network		1,741	2,532
Total	27,683	27,807	27,774

NOTE: MOS = maintenance operations squadron; AMXS = aircraft maintenance squadron; CMS = component maintenance squadron; EMS = equipment maintenance squadron.

some fraction of their primary authorized aircraft deployed and the remainder operating at the home station.[5]

Figure S.1 presents the economies of scale that demonstrate how the consolidation of backshop workloads and manpower into a small number of CRFs can achieve such reductions in manpower. The left endpoint of the curve demonstrates that, for a relatively small facility supporting a relatively small amount of flying, approximately ten manpower authorizations are required per 1,000 annual flying hours. The rightmost portion of this curve indicates that a CRF supporting a much larger workload volume is able to achieve the same levels of performance (in terms of phase throughput times and simulated not-mission-capable-for-supply rates) with significantly less manpower. This suggests that, if a repair network with a small number of relatively

---

[5] In our review of the OSD guidance and our discussions with Air Combat Command (ACC) personnel, it became apparent that the ability to conduct split operations for F-16 units is consistent with both programming guidance and recent experience.

Figure S.1
F-16 CRF Manning Requirements, Home Station

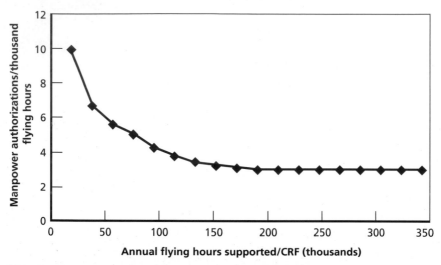

large CRFs is implemented, the total manpower requirement for these non-MG workloads can be significantly reduced.

Figure S.2 presents details on the specific CRF networks supporting CONUS aircraft that were identified by our optimization model for support of the TF F-16 fleet.[6] The two bars on the left side of the figure present the performance of the minimum-cost solution networks that have one and two CONUS CRFs. These are contrasted with two alternative networks: a maintenance network with a single CRF established at Hill AFB, and a two-CRF solution with CRFs established at Hill AFB and Robins AFB. Figure S.2 demonstrates that CONUS F-16 CRF support is somewhat insensitive to the precise number of locations that are established (it is possible to establish either one or two locations with little effect on cost performance); it also demonstrates relative insensitivity to the precise locations for CRFs. This allows for

---

[6] We also identified requirements for a CRF in each of U.S. Air Forces in Europe (USAFE) and Pacific Air Forces (PACAF); these requirements are not presented in Figure S.2 but are assumed to be constant across all CRF options for CONUS aircraft.

Figure S.2
Alternative TF CRF Solutions

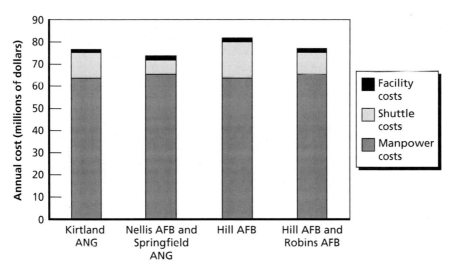

other considerations beyond the scope of this analysis to enter into the final CRF location decision. As an example, the establishment of a CRF at Hill AFB could also provide proximity to the F-16 system program office or to depot personnel.

Alternatively, if the Air Force concluded that its current F-16 maintenance operational capabilities were sufficient, our analysis identifies the potential to realize an annual savings of nearly $90 million by centralizing these backshop workloads across the TF, with no new split-operations capability created, as shown in Figure S.3. The Air Force might decide that, even though F-16 maintenance capabilities are stressed, these 1,900 backshop positions would be more effectively applied to some other career field.

The bar on the left side of Figure S.3 presents the manpower costs associated with the current system. The center bar presents the total system costs for the CRF maintenance network alternative that supports only the AD and AFRC forces, with no split-operations capability added to the CC squadrons. The bar on the right side of the figure presents the total system costs for the TF CRF network alternative,

Figure S.3
Option 2: F-16 Increased Efficiencies

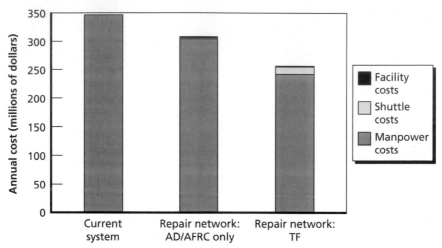

RAND *MG872-S.3*

again with no split-operations capability added to the CC squadrons. Under the current system, annual costs are $345 million, contrasted with $308 million for the AD/AFRC option ($37 million annual reduction) and $257 million for the TF option ($88 million annual reduction). Note that the manpower requirement dominates costs. The manpower cost in Figure S.3 includes both active and reserve components for the CONUS, PACAF, and USAFE CRFs, supporting both the steady-state and major combat operation surge requirements, along with those personnel who were previously in the CMS or EMS and are now reassigned to the aircraft squadron (AS), as well as the unchanged ANG phase-and-related backshop manpower for the AD/AFRC-only CRF network.

The shuttle cost associated with aircraft movement between the aircraft operating locations and the CRFs is relatively small.[7] Recent large fluctuations in the price of aviation jet fuel led us to conduct addi-

---

[7] This shuttle cost is presented only for home-station operations, because the deployed operating locations are uncertain.

tional analyses to identify how sensitive these alternative CRF network strategies would be to variations in shuttle cost.

The F-16 cost per flying hour (CPFH) used in this analysis was $6,500.[8] The CPFH includes many logistics costs in addition to aviation fuel, e.g., consumables, depot-level reparable assets, and depot maintenance costs. For the F-16C, aviation fuel constitutes $1,722, or 26 percent of the total CPFH. Because the shuttle costs are small relative to the manpower costs, the TF CRF network would be less expensive than the current system even if CPFH increased up to a factor of eight times the $6,500 value or if the price of aviation fuel increased up to a factor of 28 times the $1,722 figure (holding all other CPFH components constant).

The facility costs associated with the establishment of CRFs are also presented for the maintenance network alternatives; however, they amount to a small fraction of the total annualized costs. This suggests that, even if the facility costs computed in this analysis were understated by a factor of 10, they would not be so large as to have a material effect on the conclusions.

Of course, the Air Force could also choose to implement an alternative between enhanced effectiveness and increased efficiency for F-16 maintenance. For example, it could add a split-operations capability to some, but not all, CC squadrons. Yet another alternative for reducing manpower requirements would be to alter the deployment burden or reserve-component participation policies.

The potential for improvements in operational effectiveness and/or system efficiency exists whether the CRF network supports the TF or only the AD/AFRC forces. If the CRF network supports only the AD/AFRC forces, the associated reduction in backshop manpower is large enough to create a split-operations capability at AD and AFRC squadrons without increasing the baseline total maintenance manpower; however, not enough resources would be freed to also generate a split-

---

[8] This amount was based on U.S. Air Force, 2006, Table A4-1; the precise CPFH values given in this reference vary by F-16 series, with the F-16C, the most common series in the inventory, having a CPFH of $6,543.05 (the F-16A and F-16B had slightly lower CPFH, and the F-16D had slightly higher CPFH).

operations capability at ANG squadrons. For the increased-efficiency option, although the potential savings would be larger for the TF network, there remains an economic rationale for repair network centralization in either case.

This capability level, while broadly consistent with OSD guidance, is presented as an illustration—our analytic approach can be used to identify the resource requirements for any other capability level the Air Force deems appropriate.

KC-135

The analysis for the KC-135 (see pp. 65–90) identified similar potential for increases in effectiveness or efficiency through consolidation of certain backshop maintenance workloads into a flexible maintenance network support concept, by applying the existing Air Mobility Command (AMC) forward operating location (FOL)/regional maintenance facility (RMF) concept that is currently used to provide maintenance support to deployed forces to home-station operations as well. For the purposes of illustration, we assumed that the desired capability was the support of (1) a steady-state deployment of 40 percent of the combat direct support (CA) KC-135 fleet into two theaters for an indefinite duration and (2) a surge deployment of 100 percent of the same fleet into two theaters.

As with the F-16, our analysis identifies alternatives for KC-135 wing-level maintenance that satisfy the capability objective. Table S.2 presents an alternative that enhances the capability of KC-135 maintenance units by transferring 2,400 positions out of backshop maintenance,[9] made possible by consolidation of RMF workloads, and moving them into MG maintenance, giving each CA squadron the ability to conduct split operations. Alternatively, if the Air Force concluded that its current KC-135 maintenance operational capabilities were sufficient, it would be possible to realize an annual savings of $100 million by centralizing these backshop workloads across the TF, with no new split-operations capability created, as shown in Figure S.4.

---

[9] In Table S.2, these positions appear in the shaded cell in the "TF Repair Network" column.

Table S.2
Option 1: KC-135 Increased Operational Effectiveness

Operation	Manpower Authorization		
	Current System	AD/AFRC-Only Repair Network	TF Repair Network
Group and MOS: FY 2008 UMD	1,542	1,542	1,542
AMXS			
FY 2008 UMD	4,622	1,343	
UTC-based AMXS		2,792	4,833
UTC-based moved from MXS		741	1,366
Split operations plus-up		1,213	2,366
MXS			
FY 2008 UMD	5,573	3,351	
CRF Network		876	1,160
Total	11,737	11,858	11,267

NOTE: MXS = maintenance squadron.

Figure S.4
Option 2: KC-135 Increased Efficiencies

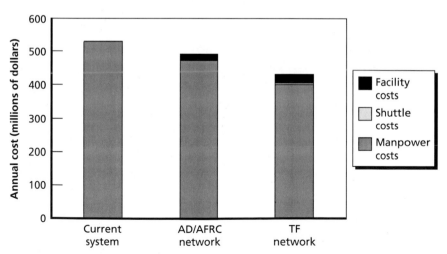

RAND *MG872-S.4*

A range of alternatives between these two endpoints also exists, as was the case in the F-16 analysis. As with the F-16, the total costs are dominated by the manpower requirement.

The bar on the left side of Figure S.4 presents the manpower costs associated with the current system. The center bar presents the total system costs for the CRF maintenance network alternative that supports only the AD and AFRC forces, with no split-operations capability added to the CA squadrons. The bar on the right side of the figure presents the total system costs for the TF CRF network alternative, again with no split-operations capability added to the CA squadrons. Under the current system, annual costs are $531 million, contrasted with $488 million for the AD/AFRC option ($43 million annual reduction) and $429 million for the total force option ($102 million annual reduction).

The manpower cost presented here includes both active and reserve components for all AMXS, MXS, AS, and CRF positions capable of supporting both the steady-state and major combat operation scenarios considered. There is a small shuttle cost associated with aircraft movement between the aircraft operating locations and the CRFs.[10] As we did for the F-16, we conducted additional analyses to identify how sensitive these alternative KC-135 CRF network strategies were to variations in the shuttle cost and found that the TF CRF network alternative would be less expensive than the current system even if the CPFH increased up to a factor of 27 times its business year 2008 level or, holding all other CPFH components constant, if the price of aviation fuel increased up to a factor of 43 times its assumed level within the CPFH. Once again, facility construction costs constitute a relatively small fraction of the total system costs.

The KC-135 CRF network concept also offers potential benefits whether the network supports only the AD/AFRC forces or the TF. An AD/AFRC CRF network generates backshop manpower reductions sufficient to create a split-operations capability at AD and AFRC squadrons without increasing the baseline maintenance manpower,

---

[10] This shuttle cost is presented only for home-station operations because of the uncertainty associated with deployed operating locations.

but it would not achieve sufficient reductions to also generate a split-operations capability at ANG squadrons. On the other hand, if the concepts are applied to the TF, there are sufficient backshop manpower reductions to create split-operations capabilities for all Air Force units, i.e., AD, AFRC, and ANG units. Were the focus instead placed on increased efficiency, the potential savings associated with the TF CRF network would be larger than the savings achieved if only AD/AFRC resources were rebalanced with the network; however, an economic rationale for repair network centralization exists in either case.

A broader view should also consider options for rebalancing resources across mission design series to meet the most pressing needs of the future security environment. Similarly, rebalancing options should also consider the reprogramming of resources between maintenance and other career fields, given projections of relative levels of future demand. Review and assessments of OSD guidance, such as the Steady-State Security Posture, could be used to help the Air Force make such determinations among aircraft and across career fields.

Acknowledgments

Many people inside and outside the Air Force provided valuable assistance and support to our work.[11] We thank Lt Gen (Ret) Kevin Sullivan, AF/A4/7, and Lt Gen Terry Gabreski, AFMC/CV, who sponsored this research and continued to support it through all phases of the project.

We are grateful to our project's action officers, Lt Col Cheryl Minto and Lt Col David Koch, AF/A4MM, for their many contributions to this effort. On the Air Staff, we thank Maj Gen Robert McMahon, AF/A4M, Maj Gen Gary McCoy, AF/A4R, and Grover Dunn, AF/A4I, along with their staffs. Their comments and insights have sharpened this work and its presentation.

At the major commands, we thank Maj Gen David Gillett, ACC/A4, Lt Col Ronald Kieklak, ACC/A4F16, and CMSgt Timothy Tisdale, ACC/A4F16, and their staffs for assisting in our F-16 analysis. We also thank Brig Gen Kenneth Merchant, AMC/A4, and his staff (particularly Capt Jerrymar Copeland, AMC/A4MXE) for providing equally great support to our KC-135 analysis. We were also fortunate to receive tremendous assistance from Brig Gen Frank Bruno, AFMC/A4, and his staff throughout the entire research process.

At the Air National Guard Bureau (NGB), we thank Col Rich Howard, NGB/A4, and his staff (including Col Steph Dowling, NGB/A4M, Col Tom Redford, NGB/A4R, Col Dave Whipple, ANG Advisor to AF/A4M, Col Tom Murgatroyd, ANG Advisor to AMC/A4, and Col Chuck Melton, ANG Advisor to ACC/A4), Col Wayne

---

[11] All office symbols and military ranks are listed as of the time of the research.

Shanks, Chairman, ANG Fighter Council (Combat Air Forces), and Col Gary Nolan, Chairman, ANG Strategic Aircraft Maintenance Council (Mobility Air Forces), for sharing their comments and insights. We were also fortunate to have visited the Air Force Reserve Command, where we received valuable feedback from Brig Gen Elizabeth Grote, AFRC/A4, Col Faylene Wright, AFRC/A4M, and their staffs.

We visited many flying units and other organizations during the course of this study, and we received valuable information and feedback from all of them. We thank Col James Howe, 6 MXG/CC, at Mac-Dill AFB; the 388th Fighter Wing (FW), 419th FW, and F-16 system program office at Hill AFB; Maj Art Sollombarger, ANG, at the 171st ARW, Pittsburgh IAP; Col Joseph Swillum, 57 MXG/CC, at Nellis AFB; Col Algene Fryer, 56 MGX/CC, at Luke AFB; Col Martin Park, ANG, at the 161st ARW, Luke AFB; Col Evan Miller, OC-ALC/XP, at Tinker AFB; and Col James Nally, 827 ACSG/CC, at Tinker AFB. In particular, Col Joe Brandemuehl, Col Ted Metzgar, Maj Bart Van Roo, and maintenance personnel of the 115th FW, ANG, Truax Field, provided insights into the deployed maintenance concepts, deployed operations, and logistical aspects of the ANG "rainbowing" concept.

We also visited a number of deployed units, where we learned a great deal about the special challenges associated with aircraft maintenance in the deployed environment. We are especially grateful to Lt Gen Gary North, U.S. Air Forces Central (AFCENT)/CC, who sponsored the visit, and Col Peter Hunt, US AFCENT/Director of Staff, Shaw AFB SC, who helped coordinate our travel to the deployed units. We thank Col Perry Oaks and Lt Col James Bruns, 332nd EMXG, and Lt Col Robert Lepper and Capt Timothy Casey, 379th EMXG, for the candid discussions they had with us and for allowing us to visit their maintenance operations. We thank Col David Carrell AFCENT/A4 and his entire staff, including Lt Col David Yockey, MSgt Thomas Colvin, AFCENT/A4M, and Brent Phillips, AF/A4MM, for the outstanding assistance they provided, both in advance of our visit and during our travel within Southwest Asia; we also offer a special thanks to MSgt Ryan Fondulis, who accompanied the RAND team during the site visits.

We benefited from conversations with Col Brent Baker, Air Force Global Logistics Support Center/CC, at Scott AFB. Rogelio Hudson at Mobility Air Forces Logistics Support Center/LGWAF, Scott AFB, shared with us his expert knowledge of Air Force data systems and pointed us to other points of contact. Arthur Eggleton at AFMC/A4RM helped us get the weight and cube data associated with the shipping of aircraft components. John Cilento, ACC/A3TB, provided insights into the relationship between programmed training flying hours and deployed flying hours.

We received invaluable support from Richard Enz and the Reliability and Maintainability Information System (REMIS) office at Wright-Patterson AFB. They provided extensive amounts of data in a timely manner and answered several questions along the way.

Our LCOM analyses were greatly aided by the support we received from the staffs of Donald White, 2nd Manpower Requirements Squadron/MRL, at Langley AFB, and Shenita Clay, 3rd MRS/MRL, at Scott AFB.

At RAND, Candice Riley provided exceptional technical programming support to our analysis of REMIS maintenance data. We benefited greatly during our project from the inputs, comments, and constructive criticism of many RAND colleagues, including (in alphabetical order) Laura Baldwin, Natalie Crawford, Greg Hildebrandt, Nancy Moore, Patrick Mills, Don Snyder, and Donald Stevens. We thank Megan McKeever for her assistance throughout the production of this monograph. We also thank our editor, Janet DeLand, for her assistance, which greatly improved the presentation and readability of the document.

We would especially like to thank our RAND colleagues John Halliday and Dick Hillestad for their thorough reviews, and Adam Resnick for his review of the mathematical models; their comments helped shape this monograph into its final, improved form.

As always, the analysis and conclusions are the responsibility of the authors.

Abbreviations

AB	Air Base
AD	active duty
AFB	Air Force Base
ACC	Air Combat Command
AEF	Air and Space Expeditionary Force
AETC	Air Education and Training Command
AFMC	Air Force Materiel Command
AFRC	Air Force Reserve Command
AFSC	Air Force Specialty Code
AFTOC CAIG	Air Force Total Ownership Cost Cost Analysis Improvement Group
AGE	aerospace ground equipment
AMC	Air Mobility Command
AMXS	aircraft maintenance squadron
ANG	Air National Guard
AOR	area of responsibility
AS	aircraft squadron
ASD	average sortie duration

BNRTS	base not reparable this station
BRAC	base realignment and closure
BSP	Baseline Security Posture
CA	combat direct support
CC	combat-coded
CCRF	contingency centralized repair facility
CIRF	centralized intermediate repair facility
CMS	component maintenance squadron
CONOP	concept of operation
CONUS	continental United States
CPFH	cost per flying hour
CRF	centralized repair facility
DoD	Department of Defense
E&E	electrical and environmental
ECSS	Expeditionary Combat Support System
EMS	equipment maintenance squadron
ERRC	expendability, recoverability, reparability category
EUCOM	U.S. European Command
FedEx	Federal Express
FOL	forward operating location
FYDP	Future Years Defense Program
GLSC	Global Logistics Support Center
GATES	Global Air Transportation Execution System

GUARDIAN	the Air National Guard Information Analysis Network
ILM	intermediate-level maintenance
ILP	integer linear programming
IPG1	Issue Priority Group 1
JEIM	jet engine intermediate maintenance
LANTIRN	Low-Altitude Navigation and Targeting Infrared for Night
LCOM	Logistics Composite Model
LEA	Logistics Enterprise Analysis
LRU	line replaceable unit
MAF	man-hour availability factor
MAJCOM	major command
MCO	major combat operation
MDS	mission design series
MG	mission generation
MOS	maintenance operations squadron
MPES	Manpower Programming and Execution System
MRS	Manpower Requirements Squadron
MXS	maintenance squadron
NDI	nondestructive inspection
NGB	National Guard Bureau
NIIN	National Item Identification Number
NMCS	not mission capable for supply

NSN	national stock number
OCONUS	outside the continental United States
OIMDR	organizational and intermediate maintenance demand rate
OPTEMPO	operating tempo
ORG	Organizational Code
ORGT	organizational title
OSC	Organizational Structure Code
OSD	Office of the Secretary of Defense
OSTR	organizational structure
PAA	primary authorized aircraft
PA&E	Program Analysis and Evaluation
PACAF	Pacific Air Forces
PACOM	U.S. Pacific Command
PAD	Program Action Directive
PAF	Project AIR FORCE
PBD	Program Budget Decision
PDM	programmed depot maintenance
PE	periodic inspection
PEC	Program Element Code
QDR	Quadrennial Defense Review
RC	Reserve Component
RDB	Requirements Data Bank
RE21	Repair Enterprise 21

REMIS	Reliability and Maintainability Information System
RIC	Resource Identification Code and Title
RMF	regional maintenance facility
RSP	readiness spares package
SHOP	shop code
SPG	Strategic Planning Guidance
SPO	system program office
SRAN	stock record account number
SSSP	Steady-State Security Posture
TF	Total Force
TMC	Type Maintenance Code
TNMCS	total not-mission-capable for supply
TRC	technology repair center
UMD	Unit Manning Document
USAFE	U.S. Air Forces in Europe
UTC	Unit Type Code
UTE	utilization rate
WUC	Work Unit Code

CHAPTER ONE

Introduction

Background

The Air Force has developed an extensive set of logistics resources and concepts to support training, deployment, employment, and redeployment of air, space, and cyber forces. Logistics activities, such as aircraft maintenance, inventory management, and distribution of assets, directly affect the Air Force's ability to generate aircraft sorties. Furthermore, the execution and management of these logistics activities, along with the associated investments in logistics information systems, consume a large fraction of the total Air Force budget. Much of this "logistics enterprise" was designed during the Cold War era and focused on supporting the operational concepts of that time. For example, F-16 maintenance was designed around the concept of an entire squadron deploying to a single forward operating location (FOL) and executing its wartime mission, with a design objective of the squadron being self-sufficient for the initial 30 days of the deployment.

Since the establishment of the Iraqi no-fly zones following the Persian Gulf War in 1991, however, the Air Force has been continuously engaged in rotational deployments. These deployments have supported a full range of operations, from contingency operations over Serbia, Iraq, and Afghanistan, to deterrence operations, such as Operations Southern Watch and Northern Watch, to peacekeeping, to humanitarian support. These rotational deployments involving different sets of missions present challenges much different from those that were used to determine the resource requirements and support concepts that resulted in the development of the current support posture, one still largely based on assumptions developed during the Cold War. Changes to the operational environment, such as rotational deployments of less-than-squadron-size F-16 units, may warrant a change to the Air Force's logistics infrastructure and concepts of operation, but the questions of "what changes" and "to what extent" remain unanswered.

Research Motivation

Two important influences provided the motivation for this research. One is the changes in the Office of the Secretary of Defense (OSD) guidance, and the other is the substantial reduction in Air Force manpower.

Changes in Guidance

Department of Defense (DoD) Strategic Planning Guidance (SPG) and the Quadrennial Defense Review (QDR) for 2000 (U.S. Department of Defense, 2001) specify that capabilities will be created to accomplish the following, known as the 1-4-2-1 strategy: *one*, ensure homeland defense; deter aggression in *four* major areas of the world and engage in a number of small-scale contingencies, if needed; if deterrence fails in the four areas of strategic importance, be able to engage in *two* major combat operations (MCOs) simultaneously, with the ability to win *one* decisively while engaging in the other until the first is won, and then to win the second.

The 2004 SPG contains defense planning scenarios to be used for programming operational and support requirements. These include scenarios associated with MCOs, a Baseline Security Posture (BSP), homeland security (as part of the Global War on Terrorism), and small-scale contingencies. The guidance recognizes that the U.S. military is likely to be engaged in several global operations at any given time. The guidance also recognizes that MCOs, if they occur, are likely to be initiated from an already-engaged posture. This OSD guidance instructs the services to size their operational and support forces to execute two MCOs while still providing homeland security, indicating that BSP activities may be curtailed, if necessary, to meet MCO and homeland security requirements.

The 2006 DoD SPG directs the services to focus on developing the capabilities to defend the homeland, conduct irregular warfare, and conduct and win conventional campaigns. This guidance replaces the BSP with a set of Steady-State Security Posture (SSSP) scenarios.

Looking further into the future, the capabilities required of the U.S. military, along with the roles and missions for each service, may change even more dramatically than those outlined in the SPG.[1] Recent RAND Corporation analysis suggests that in the future, services may be asked to accomplish the following tasks:

- maintain a substantial and sustained level of effort to suppress terrorist and insurgent groups abroad
- support "hands-on" efforts to train, equip, advise, and assist the forces of nations that seek to suppress insurgents in their own territories
- provide support to defeat internal threats and shore up regional security to cope with external enemies
- overcome modern antiaccess weapons and such methods as theater ballistic missiles and cruise missiles (Hoehn et al., 2007).

The RAND analysis also postulates how redefining roles and missions for the services and rethinking planning requirements may better prepare each service to respond in the future. For example, with resources limited, should ground forces focus on stability operations, while the Air Force and Navy focus on large-scale power-projection operations?

Reductions in Resources

OSD guidance provides a glimpse into what future operations may require; however, fiscal constraints limit the resources available to meet these requirements. Between FY 2005 and FY 2008, more than 30,000 positions were cut from the Air Force's congressionally mandated end-

---

[1] See, e.g., U.S. House of Representatives Committee on Armed Services: Panel on Roles and Missions, 2008.

strength ceiling.[2] These manpower reductions must be achieved without sacrificing the operational capabilities outlined in DoD and Air Force planning guidance. Attrition and manpower savings achieved through base realignment and closure (BRAC) will provide some of these reductions. However, under current force employment practices, these manpower reductions may leave the active component without sufficient end-strength authorizations to support some operational requirements.

The Air National Guard (ANG), while less significantly affected by PBD 720-related manpower reductions than the active duty (AD) force, is still affected by force structure changes in support of the QDR and BRAC. A significant number of legacy aircraft will be retired (U.S. Department of Defense, 2005a),[3] many of which are in the ANG. Under current force employment practices, force structure reductions will not affect ANG end-strength manpower authorizations, but they may leave the ANG without sufficient clearly defined missions to support current operational requirements to employ their existing end strength.

In the past, mandated manpower reductions have led to the transfer of mission assignments to contractors. However, current manpower reductions (in PBD 720) also reduce contractor support. Thus, an increase in contractor support cannot be used to solve the problem of maintaining operational effectiveness with reduced resources.

---

[2] The Ronald W. Reagan National Defense Authorization Act for Fiscal Year 2005 authorized an Air Force end strength of 359,700; the National Defense Authorization Act for Fiscal Year 2008 authorized an Air Force end strength of 329,563. The Air Force had planned for a further reduction of 13,000 positions in the U.S. Air Force FY 2009 President's Budget Request. Program Budget Decision (PBD) 720 (U.S. Department of Defense, 2005b) directed additional manpower reductions resulting in a total reduction of approximately 57,000 personnel through FY 2011. Note, however, that Secretary of Defense Robert Gates announced in June 2008 that additional manpower cuts below the FY 2008 end strength were to be put on hold.

[3] For example, the BRAC Commission calls for the elimination of the flying mission of a number of ANG flying units operating A-10, F-16, C-130, and KC-135 aircraft.

Purpose, Objectives, and Approach of This Analysis

Recognizing the importance of the logistics enterprise to the execution of its larger set of roles and missions, the Air Force has initiated a set of interrelated transformation activities, including Air Force Smart Operations for the 21st Century and Expeditionary Logistics for the 21st Century,[4] in an attempt to ensure that logistics capabilities evolve in a manner and at a rate compatible with the ongoing evolution of the operational environment. These large-scale activities comprise a broad set of initiatives, including Repair Enterprise 21 (RE21), Centralized Asset Management, Expeditionary Combat Support System, Global Logistics Support Center (GLSC), Air Force Maintenance for the 21st Century, depot lean actions, and purchasing and supply-chain management analyses; all are intended to improve the effectiveness and efficiency of Air Force logistics activities.[5] Each initiative appears to have merit, and each may make a positive and incremental contribution to improved logistics performance. However, ensuring that these efforts (which may at times promote conflicting objectives) remain integrated into a larger vision for transformation is a cause for concern.

In light of these changes to the Air Force's operational environment and ongoing transformational initiatives, senior Air Force leaders requested that RAND Project AIR FORCE (PAF) undertake a comprehensive strategic reassessment of the entire Air Force logistics enterprise—to reidentify and rethink the basic issues and premises on which the Air Force plans, organizes, and operates its logistics enterprise.

An effective logistics enterprise strategy will support the evolving requirements for expeditionary deployment and employment of Air Force assets—including assets residing in the Air Force Reserve Com-

---

[4] Air Force Smart Operations for the 21st Century is a method for institutionalizing continuous process improvement in the entire Air Force (including, but not limited to, logistics); Expeditionary Logistics for the 21st Century is an effort to transform current logistics processes to provide better support to the warfighter.

[5] Further information on many of these initiatives can be found in U.S. Air Force, Deputy Chief of Staff for Installations and Logistics Directorate of Transformation (AF/A4I) (2007).

mand (AFRC) and ANG. At a fundamental level, the logistics enterprise strategy must answer the three questions listed below. Accordingly, this research project addresses each.

1. **What will the logistics workload be?** Since logistics is a demand-driven process, an effective enterprise strategy must rest on a sound understanding of future force structure and levels of activity, the environmental factors that will influence future workloads and requirements, and especially the evolving requirements for expeditionary deployment and employment of Air Force assets from a joint and coalition-partner viewpoint.

2. **How should the logistics workload be accomplished?** That is, how should the enterprise be structured and organized to handle the required logistics workloads? The strategic decisions made here should recognize that key management options and important resource trade-offs occur in the following areas:

    • **"Stockage" solutions versus "response" solutions.** Support of deployed and engaged forces has historically involved a blend of stocks (readiness spares packages, war readiness engines, prepositioned war reserve materiel, etc.) and responsive support (resupply, deployed intermediate-level maintenance [ILM], centralized intermediate repair facility [CIRF] support, etc.). Strategic issues here involve the ability to forecast usage rates and requirements, in-theater footprint issues, and unit "ownership" versus "sharing" of assets.

    • **Local maintenance versus network maintenance.** Flexibility exists with respect to the location of maintenance activities that are not directly tied to sortie generation and recovery. For example, off-equipment component repair can be conducted on base in backshops,[6] or off base at Air Force Materiel Command (AFMC) depot facilities, at a

---

[6] *Backshops* refers to one of three levels of Air Force aircraft maintenance. Historically, backshops perform intermediate-level maintenance; depots perform the highest level; and organizational, or "flightline," the lowest.

centralized repair facility (CRF), or at contractor facilities. This geographic dimension has important strategic considerations, such as the extent of reliance on transportation and the potential for realizing economies of scale in maintenance operations. Deployment and distribution requirements play a key role in determining which option is preferable for a specific maintenance action. Special consideration should be given to the differing risks associated with "self-sufficient" versus "network" maintenance postures.

- **Contract versus organic maintenance.** The Air Force can partner with industry to handle component repair, aircraft inspection, end-item overhaul, aircraft programmed depot maintenance (PDM), and modifications workloads. It can also choose more comprehensive contract logistics support arrangements. This set of "outsourcing" decisions can have strategic flexibility and effectiveness implications as well as efficiency effects.

- **Commodity orientation versus weapon system orientation.** The Air Force has historically blended these approaches. Technology repair centers (TRCs) and commodity councils focus on the economies of scope and scale associated with grouping like assets, while system program offices (SPOs) capitalize on the effectiveness trade-offs that can be achieved by focusing on the weapon system as a whole. These organizing constructs have sophisticated strategic nuances, such as the efficiency of TRCs compared with the simplicity of managing weapon-system-unique maintenance networks.

3. **How should these questions be revisited over time?** To maintain currency, the Air Force logistics community should periodically readdress all these questions. The assignment of such responsibilities and the development of the necessary analytic capabilities to support such decisionmaking need to be consid-

ered to ensure that the logistics enterprise remains dynamic and responsive to changes in the strategic environment.

The allocation of logistics capabilities along the dimensions described above will define the strategic design of the Air Force's logistics enterprise. But how should these allocations be made? That is, what strategic rationale should be used to make each of these decisions? Consider the decision as to whether a specific component, for example, an F-16 line replaceable unit (LRU), should be repaired at a contractor facility or at an AFMC depot. Should this decision hinge primarily on cost, or are there other strategic issues to be considered? If there are—and we believe this to be the case—what calculus should be used to make this decision?

To answer these questions, we defined the following four tasks for the RAND Logistics Enterprise Analysis (LEA) project:

- a review of OSD programming guidance to determine projected logistics system workloads
- a structured analysis of scheduled and unscheduled maintenance tasks to strategically rebalance workloads among operating units and supporting network nodes
- a strategic reevaluation of the objectives and proper roles of contract and organic support in the logistics enterprise
- a top-down review of the management of logistics transformation initiatives to ensure that they are aligned with broader logistics objectives.

This monograph addresses only a part of the broader LEA project, presenting results from analyses of the first two tasks for the F-16 and KC-135 fleets. Subsequent publications will address the other tasks and present enterprise rebalancing analyses for other weapon systems.

More specifically, this monograph details the analysis we performed to identify alternatives for rebalancing aircraft maintenance capabilities between unit-specific and network sources of repair for the F-16 and KC-135. Alternatives for the allocation of these workloads between the AD and Reserve Components (RCs) are examined in this analysis; a further examination of potential allocations to government civilians and contractors may be addressed in subsequent publications. We focus on wing-level maintenance tasks, including sortie launch and recovery workloads, aircraft inspections, on-equipment maintenance to support removal and replacement of aircraft components, shop repair of replaceable components, and time-change technical orders. The analysis does not address depot-level workloads such as PDM or overhaul of major items. We selected the F-16 and KC-135 mission design series (MDSs) because of their dissimilarities in both logistics support requirements (e.g., F-16s have a relatively short phase-inspection interval and a limited flying range, while KC-135s have a relatively long inspection interval and a much longer flying range) and the projected demands for these aircraft in support of future deployment scenarios (F-16s would be used in a smaller set of potential operations than would KC-135s, although the much larger F-16 fleet might make greater demands on the logistics system for certain operations).

The analytic approach we took to address these research questions is as follows. We begin by reviewing the defense planning guidance, translating that into a range of potential logistics workloads. Then, for each MDS, we identify what we term "aircraft squadron (AS) maintenance," that is to say, the maintenance tasks that necessarily must remain at the aircraft's operating location. We next identify the manpower requirements to support the AS workloads, both in "traditional-operations" and "split-operations" environments. All other maintenance activities are considered candidates for off-site support at repair network facilities that underpin the AS. We evaluate the performance of this maintenance network, focusing on the repair capacity, inventory, and distribution system requirements necessary to meet operational objectives. An optimization approach is utilized to design the maintenance network, identifying the number and location of network maintenance facilities. Information-system requirements, while important to the performance of the overall logistics system, are not addressed in this monograph. We conclude the analysis by identifying alternatives for rebalancing the resources invested in the AS versus those invested in the maintenance network and presenting a range of options lying along the tradespace of "enhancing operational effectiveness" to "extracting savings from the logistics enterprise."

Our analyses address the costs and benefits of an enterprise approach that would be configured to support the Total Force (TF). We also evaluate an enterprise option that would be used to support only AD and AFRC forces.

How This Report Is Organized

The report has five chapters. Chapter Two details the identification of logistics workloads from planning and programming guidance. Chapter Three presents the results of our analysis for the F-16, while Chapter Four does the same for the KC-135. Chapter Five contains our research conclusions. The report also has six appendixes:

A. Maintenance Manpower Authorizations
B. Modeling F-16 Maintenance with the Logistics Composite Model
C. Analysis of Phase and Periodic Inspection Maintenance Using the Reliability and Maintainability Information System (REMIS)
D. Integer Linear Programming Model
E. Estimating KC-135R Maintenance Manpower Requirements
F. Estimating CRF Component-Repair Pipeline Effects.

These appendixes contain detailed information about how various calculations to support the analysis were made. Cross-references in the text indicate where that information applies.

CHAPTER TWO

Projecting Logistics System Workloads

The first question a logistics enterprise analysis needs to address is, What will the logistics workload be? Logistics is a demand-driven process, and future demands will be generated by both more-predictable home-station training operations and less-predictable deployment and employment operations. Current OSD guidance suggests that the demands for deploying air power are likely to maintain the current high level of operating tempo (OPTEMPO) for the foreseeable future. The Air Force must be prepared to support a wide range of operational demands in locations that are difficult to anticipate. While the Air Force still must ready itself for major conflicts, the nature of those engagements is likely to differ dramatically from the scenarios envisioned in the past. As recent history has illustrated, the Air Force must also remain ready to support national interests in other types of operations, ranging from peacekeeping, to humanitarian assistance, to support of special operations.

Because the analysis detailed in this monograph is focused on wing-level aircraft maintenance, the key data we used to identify future workloads are the projected number of aircraft operating out of each location (whether at home station or deployed) and the total number of flying hours at each aircraft operating location. Aircraft flying hours are commonly used as a predictor variable to estimate such maintenance workload drivers as engine failures and aircraft phase inspections.

We identified the home-station beddown of all U.S. Air Force aircraft, as of the end of FY 2008. For the F-16, the fleet size was identified as 1,074 total primary authorized aircraft (PAA), of which 747 were combat coded (CC). For the KC-135, we identified a fleet size of 419 PAA, of which 394 were combat direct support (CA). We based training flying-hour requirements on the FY 2008 POM Flying Hour Program provided by the Air Force Total Ownership Cost Cost Analysis Improvement Group (AFTOC CAIG). We reviewed OSD Program Analysis and Evaluation (PA&E) guidance documents to identify potential Air Force deployment and employment requirements. An important direction from these documents is that the services should plan to commence MCOs from a posture in which a significant fraction of the force is already deployed in support of "lesser contingencies." These planning documents present alternatives for force requirements in both the "steady state" of continuous small-scale deployments (the BSP[1]) and in support of MCOs. Where these documents gave specific aircraft operating locations, we used those sites; for deployments that did not give specific operating locations, we selected sites based on a review of the Mobility Capabilities Study (U.S. Department of Defense and the Joint Chiefs of Staff, 2005), which addressed this same set of deployments. We based flying-hour requirements on planning factors presented in the Air Force War and Mobilization Plan (U.S. Air Force, 2000), using our judgment to adjust the flying hours for different types of deployments (e.g., humanitarian relief operations).

Figure 2.1 depicts a notional, but representative, example of how we translated the operational requirements specified in these documents into logistics workloads. The horizontal axis presents time, in years across the Future Years Defense Program (FYDP). Plotted on the vertical axis are the flying hours projected to be generated over that interval for a notional aircraft MDS. The sum of the "training" and "BSP: training offset" areas equals the training flying-hour requirement. Note, however, that only the fraction identified as training is projected to be flown at the aircraft's home station. This is due to the realization that, in a continuously deployed environment, some of the aircraft that were projected to be training at the home station are deployed in support of BSP operations. The "BSP: training offset" area captures those

---

[1] The BSP has since been replaced by the SSSP.

Figure 2.1
Notional Flying Hour Requirements

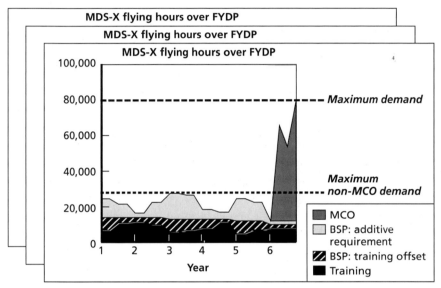

flying hours that were projected as training requirements but that must instead be flown in support of BSP-type operations.[2] The sum of the "BSP: training offset" and "BSP: additive requirement" areas is equal to the total BSP flying-hour requirement. The extent to which the "BSP: additive requirement" region is larger than the "BSP: training offset" region (at any point in time) indicates the relative difference in OPTEMPO between training and BSP flying. While recognizing that there is a steady-state deployment requirement that must be supported, the services are also expected to have the capability to conduct traditional MCOs. The "MCO" curve shows the dramatic increase in flying-hour requirements associated with conducting such operations.

---

[2] It is assumed that the missions flown during a BSP deployment are perfectly substitutable for training flying requirements. While it may not be the case for all MDSs, we make the assumption that these hours are a pure "offset" to training, based on September 2007 discussions with John Cilento, ACC/A3TB.

The OSD guidance is also clear that the services should not plan for any single future. Figure 2.1 attempts to present this concept by showing multiple flying-hour requirement charts for this MDS. In this analysis, we present a method for evaluating ranges of possible futures consistent with the guidance documents, identifying what those alternative futures imply in terms of maintenance manpower requirements.

CHAPTER THREE

Alternatives for Rebalancing F-16 Maintenance Resources

This chapter details how we determined the maintenance tasks and manpower requirements for the F-16 aircraft. We first lay out the scope of the manpower analysis and then describe how we calculated the maintenance workload. We follow that with a description of how we identified the staffing levels for various work centers. The next section provides network options for the TF, and the subsequent section does the same for only the AD and AFRC. The last section of the chapter presents our conclusions.

Scope of the Manpower Analysis

Table 3.1 shows the end-FY 2008 manpower authorizations associated with F-16 wing-level maintenance, as presented in Unit Manning Documents (UMDs), across the AD, ANG, and AFRC.[1] This set of

---

[1] In May 2008, Gen Michael Moseley, Chief of Staff of the Air Force, approved Program Action Directive (PAD) 08-01. This directive would realign bomber, rescue, and fighter (including F-16) aircraft maintenance units into their attendant flying squadrons and transfer all their remaining maintenance functions into a new materiel group. This organizational change was scheduled to be completed by November 2008, with initial application to the AD and AFRC. The implementation of PAD 08-01 was placed on hold in June 2008, following Secretary of Defense Robert Gates's recommendation for Gen Norton Schwartz to serve as the 19th Chief of Staff of the Air Force. While our research team provided information to the Air Force team that was tasked with development of the PAD 08-01 reorganization, the mission-generation (MG) concept presented in this monograph could be viewed as an alter-

Table 3.1
F-16 Maintenance Manpower Authorizations

| | | Manpower Authorization | | | | |
| | | ANG | | AFRC | | |
Operation	AD	Part-Time	Full-Time	Part-Time	Full-Time	Total
Group and MOS	1,954	598	631	82	98	3,363
AMXS	6,147	2,628	1,674	413	281	11,143
CMS and EMS						
Propulsion and avionics	1,516	480	693	68	106	2,863
AGE and munitions	2,539	936	363	176	79	4,093
Phase and related	3,143	1,481	1,233	196	168	6,221
Total	15,299	6,123	4,594	935	732	27,683

SOURCE: F-16 FY 2008 UMD.

NOTE: This table presents manpower authorizations, which typically exceed the actual manpower assigned to any organization. Because this analysis focuses on maintenance manpower requirements, we present all manpower in terms of authorization levels.

manpower authorizations comprises approximately 28,000 positions, located in the following organizations: maintenance group, maintenance operations squadron (MOS), aircraft maintenance squadron (AMXS), component maintenance squadron (CMS), and equipment maintenance squadron (EMS).[2] This analysis does not address the approximately 3,400 supervisory and support positions located in the maintenance group and MOS.

Organizational-level, or "flightline," maintenance is primarily performed by the AMXS, while F-16 intermediate-level, or "backshop," maintenance is primarily performed by the CMS and EMS. This analysis for the F-16 focuses on alternatives for rebalancing the resources invested in the AMXS, CMS, and EMS. Within the CMS, we exclude propulsion, or jet engine intermediate maintenance (JEIM), and avionics maintenance because these workloads are in the process of being

---

native for maintenance reorganization, extending beyond the PAD 08-01 realignments but maintaining a separate maintenance organization.

[2] See Appendix A for more details regarding this manpower analysis.

removed from wing-level organizations and centralized under RE21 initiatives; these maintenance "shops" account for approximately 3,000 of the remaining manpower positions. We further exclude aerospace ground equipment (AGE) and munitions maintenance workloads, for technical reasons to be discussed in detail later in this chapter,[3] setting aside a further 4,000 manpower authorizations. The remaining CMS and EMS shops account for roughly 6,000 manpower positions. The largest workload performed by these shops is associated with aircraft phase inspections. This CMS and EMS remainder, when added to the 11,000 AMXS manpower authorizations, defines the scope of this analysis, as indicated by the cells highlighted in gray in Table 3.1.

This analysis identifies alternatives for rebalancing resources within these two areas, with an objective of defining endpoints on a cost-capability trade-off curve. At one endpoint, we identify a posture that maintains the current level of logistics resources but increases Air Force capability; at the other endpoint, we present a solution that maintains the current level of capability at a reduced level of resource investment.

A desirable aircraft maintenance posture depends on the allocation of workload between the sortie-generating capability (which must remain at the aircraft's operating location) and the network capability (which could potentially be provided anywhere). To begin, we must first specify those maintenance tasks or workloads that are not candidates for removal from the squadron, that is, those tasks that are technically impossible to remove from the aircraft's operating location. We use the term *AS maintenance* to denote such workloads. The policy option that we evaluate in this report is the removal of *all* other tasks, with the responsibility for all non-AS workloads being assigned to the maintenance network.

---

[3] We may revisit the AGE and munitions maintenance manpower in a future LEA research effort.

Maintenance Workload Analysis

For the F-16, we used the Logistics Composite Model (LCOM)[4] to analyze the maintenance workload. LCOM is a discrete-event Monte Carlo simulation that models the aircraft sortie-generation process and identifies the logistics resources (primarily maintenance personnel and spare parts) necessary to maintain operational aircraft. The LCOM simulation enabled us to aggregate workloads to the maintenance-shop level and to identify each shop's workload distribution across three categories. The first category is MG, including removal and replacement of failed LRUs; these tasks must be accomplished at the aircraft's operating location. The second category is component repair workloads that could potentially be performed away from the AS. As noted above, the RE21 initiatives are already implementing F-16 propulsion and avionics component repair at sites removed from the aircraft's operating location. The third category of workload includes phase-related tasks. *Phase-related* refers not solely to those specific tasks that appear on the phase work cards, but also to all the maintenance that is accomplished during the phase process. Phase-related workloads could also be accomplished away from the aircraft's operating location. A key distinction is that, while component repair entails transporting broken components between aircraft operating locations and a repair facility, moving phase-related work involves flying the aircraft from its operating location to a CRF.

Table 3.2 presents the results of the LCOM analysis, identifying the distribution of these shops' workload among MG, phase-related, and component repair. We first ran the standard LCOM model for a "wartime sustained" scenario for the F-16 Block 40.[5] We then reran the model for the same scenario, but with the logic modified

---

[4] See Appendix B for an explanation of the LCOM model.

[5] While we performed LCOM simulation runs for other F-16 blocks, all the analysis presented in this monograph is based on simulation runs of the F-16 Block 40. For those shops whose workload differs dramatically across different blocks, this approach may yield highly inaccurate results. However, our understanding is that the primary maintenance difference with respect to block number is associated with avionics maintenance, which, as mentioned before, has been excluded from this analysis.

Table 3.2
Groupings of F-16 Shops

Shop	Percentage of Workload		
	Phase	Component Repair	On-Equipment Nonphase
Primarily flightline support			
Flightline crew chief	6	0	94
Flightline E&E	3	5	92
Flightline attack control	7	0	93
Flightline engines	0	1	99
Weapon loaders	0	0	100
Weapon maintenance	28	0	72
Mixture of flightline and backshop workloads			
Metals technology	18	40	42
Egress	33	15	53
Fuels	34	1	66
NDI	5	59	36
Structural repair	7	26	68
Mixture of phase-related and component repair			
Aircraft inspection	100	0	0
Pneudraulic	24	76	0
Armament	20	74	6
E&E	33	60	7
Component repair workload			
Wheel and tire	0	100	0
Survival equipment	0	100	0
JEIM	0	100	0
Engine accessories	0	100	0
Engine test cell	0	100	0
Sensor/LANTIRN	0	100	0
Avionics test stations	0	100	0
Electronic warfare	0	100	0

NOTE: E&E = electrical and environmental; NDI = nondestructive inspection;
LANTIRN = Low-Altitude Navigation and Targeting Infrared for Night.

such that no aircraft were inducted into phase. For each shop, we identified the change in workload between the two-phase induction rules;
the difference then defined the "phase-related" workloads that are presented in the leftmost data column of Table 3.2. We classified all work-

load other than phase identified in LCOM as off-equipment into the component repair category. All remaining workload was classified into the rightmost data column, "On-Equipment Nonphase"; these are the MG workloads discussed earlier. Appendix B provides further detail on the analytic process we developed to classify each shop's workload, using LCOM. As mentioned previously, neither AGE nor munitions flight maintenance workloads are addressed in this monograph.[6]

Considering the shops in the aggregate, we separate them into four categories based on the workload analysis: primarily flightline, mixture of flightline and backshop, mixture of phase-related and component repair, and component repair. The six shops listed at the top of Table 3.2 are those that are currently in the AMXS. These shops provide primarily flightline support. The preponderance of their workload appears in the "On-Equipment Nonphase Workloads" column and includes the tasks that are associated with sortie generation and removal and replacement of LRUs. Note that the phase column indicates that many of these shops provide some level of support to the phase operations. Since the policy option evaluated in this monograph entails the removal of all non-MG workloads, including phase, from the aircraft's operating site, maintenance personnel at the CRF will be responsible for performing those workloads. In fact, nearly 30 percent of the weapon maintenance shop's workload consists of phase-related tasks. However, none of the other five AMXS shops have more than 8 percent of the workloads in the phase-related and component repair categories.

The personnel manning the flightline crew chief, flightline E&E, and flightline engines shops have the same Air Force Specialty Code (AFSC)[7] as those in the aircraft inspection (2A5X1), E&E (2A6X6), and JEIM (2A6X1), respectively. Thus, we could reassign all these non-MG tasks to the corresponding non-AMXS shops, since anyone at the non-AMXS shop would be qualified to perform any tasks associated with his or her AFSC. This assumes that the AMXS shops do

---

[6] LCOM does not simulate AGE maintenance activity. Further, a different model, the Munitions Assessment LCOM Tool, is used to simulate munitions flight workload; thus, munitions workloads were also not evaluated in this analysis.

[7] An AFSC describes the job classification of a manpower position.

not have any specialized equipment necessary to perform these tasks that is not found at the non-AMXS shops. For flightline crew chiefs, all non-MG workload is associated with phase; thus, all these tasks could be reassigned to the aircraft inspection shop. For flightline E&E, although workloads are associated with both phase-related and component repair tasks, this workload could be assigned to the E&E shop if it were assumed that this shop would perform both phase-related and component repair workloads. For flightline engines, the small amount of component repair workload could be assigned to the JEIM, even though the JEIM would not necessarily be located at the aircraft phase site. Since it does not perform any phase-related workload (including the reassigned workload from the flightline engine shop), this is not a concern.

The one problematic shop is the flightline attack control shop. We assumed that its phase-related workload could be assigned to the avionics test station shop, even though personnel with different AFSCs staff these two shops. Furthermore, because the flightline attack control shop's non-MG workload consists of phase-related tasks, these workloads would need to be assigned to a shop that is collocated with the phase facility (unless it were assumed that these workloads, while phase-related, were off-equipment tasks that could be performed at a site remote from the phase facility, allowing for transportation of avionics assets between the two sites). Because of the assumption that was made with respect to the flightline attack control workload, it is likely that some additional personnel would need to be added to the phase location to perform these tasks, although we emphasize that this is a small part of the workload (7 percent) for a fairly small shop (29 positions per 24-PAA squadron at Hill Air Force Base (AFB), according to an Air Combat Command (ACC) LCOM report (U.S. Air Force, 2003). We reassigned these tasks into our "new" LCOM task networks that were used to identify maintenance manpower requirements under the new maintenance organization. For more details, see Appendix B.

The next five shops in Table 3.2 (metals technology, egress, fuels, NDI, and structural repair) have a substantial amount of workload appearing in the "On-Equipment Nonphase" category (which must be performed at the aircraft's operating site) and summed across the phase-related and component repair categories. Because of the policy assumption in this analysis that all non-MG workloads would be assigned to a CRF, the unique AFSCs associated with each of these shops must be split. For example, the metals technology shop must retain some fraction of the shop and its personnel at the aircraft operating location to accommodate its MG workloads; we thus split this shop and reassigned the component repair and phase-related portion of its workload to the CRF. We also assume that the weapon maintenance shop is similarly split because of the distribution of its workloads. Because the maintenance manpower at each shop will have to be split in accordance with this workload division, some diseconomies will occur, since a disproportionately large fraction of the manpower must remain at the aircraft operating location to perform the sortie-generation tasks (primarily due to minimum-crew-size effects), although the benefits of centralizing the other workloads could potentially outweigh these manpower diseconomies. There may also be additional equipment requirements if each shop has specialized equipment that must remain at each operating location (as is the case today) and if there is a need for additional equipment at the CRF.

The next four shops (aircraft inspection, pneudraulic, armament, and E&E) constitute the third category of shops, those whose workloads are primarily included in the "Phase-Related" and "Component Repair" columns. These shops can be moved in their entirety to a CRF, since they provide very little direct sortie-generation support. We reassign the sortie-generation workloads of the two exceptions— the armament and E&E shops—to AMXS shops that are staffed by similar AFSCs, in this case the weapon maintenance and flightline E&E shops, respectively (in effect, the opposite of what we did for the AMXS shops earlier).

The remaining shops fall into the fourth category, those whose workloads consist entirely of component repair, as indicated by a 100 percent value in the center data column of Table 3.2. Recall that the propulsion and avionics shops are excluded from this analysis, eliminating six of the eight shops in the category from our consideration. Also, since these LCOM models were built, responsibility for survival equipment maintenance has been moved out of the maintenance organization and into the operations organization. Thus, wheel and tire is the only shop in the "purely component repair" category that is included in this analysis.

The key trade-off to be evaluated is, Are the efficiencies that can be gained through centralization of the entirety of these third-category shops and some fraction of the second-category shops large enough to offset the cost associated with splitting some shops and transporting the aircraft and components between aircraft operating locations and CRFs?

Using the logic described above to apportion the workload, we identified the work centers that would be associated with AS maintenance, as presented on the left side of Figure 3.1, and those that would be associated with a CRF, as presented on the right side of the figure.

Figure 3.1
Shop Workload Distribution Between AS and CRF Work Centers

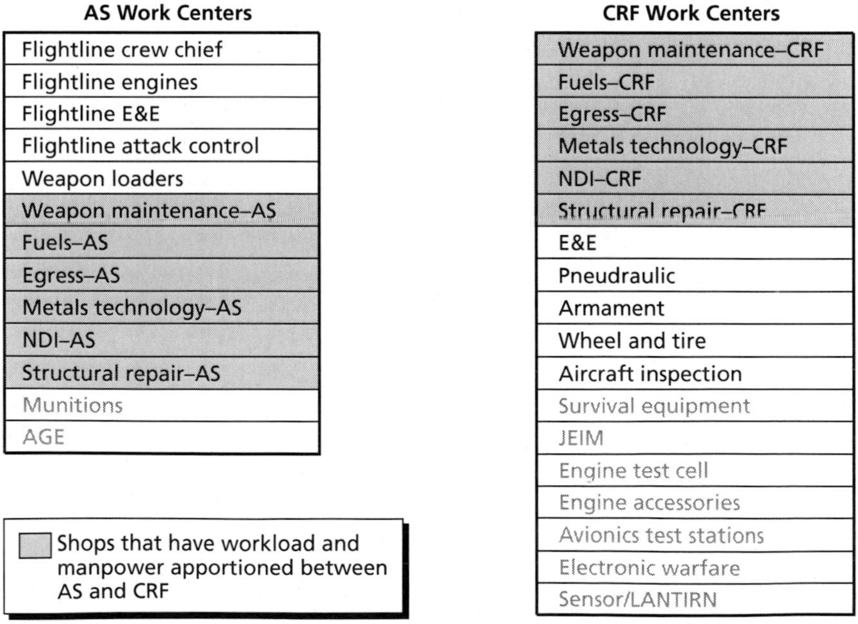

AS Work Centers	CRF Work Centers
Flightline crew chief	Weapon maintenance–CRF
Flightline engines	Fuels–CRF
Flightline E&E	Egress–CRF
Flightline attack control	Metals technology–CRF
Weapon loaders	NDI–CRF
Weapon maintenance–AS	Structural repair–CRF
Fuels–AS	E&E
Egress–AS	Pneudraulic
Metals technology–AS	Armament
NDI–AS	Wheel and tire
Structural repair–AS	Aircraft inspection
Munitions	Survival equipment
AGE	JEIM
	Engine test cell
	Engine accessories
	Avionics test stations
	Electronic warfare
	Sensor/LANTIRN

☐ Shops that have workload and manpower apportioned between AS and CRF

RAND MG872-3.1

Appearing at the top of the "AS Work Centers" column are the existing AMXS shops. Those that have "–AS" following their name are the split shops. That is to say, there is some fuels capability at the AS, and there is also some fuels capability within the CRF. Those shops that have been grayed out are excluded from the current analysis for reasons discussed previously.

Determining Staffing Levels for Work Centers

What is now needed is a method for sizing the maintenance manpower necessary to staff these work centers. Let us first consider the AS shops. We began by identifying the current manpower authorizations for the AMXS shops (flightline crew chief, flightline engines, flightline E&E, flightline attack control, weapon loaders, weapon maintenance) as presented in UMD data (see Appendix A for further details). Observe that, for a representative 24-PAA CC squadron, this entails about 220 positions (first data column of Table 3.3, "Current UMD").

Table 3.3
Split-Operations Requirements for AS Shops

| | Manpower Authorization | | |
| | | | |
Shop	Current UMD	EMS/CMS Movement	Split-Operations Plus-Up
Flightline crew chief	84	—	0
Flightline engines	15	—	9
Flightline E&E	12	—	3
Flightline attack control	11	—	7
Weapon loaders	48	—	0
Weapon maintenance	16	—	7
Supervision/support	30	—	6
Fuels–AS	—	12	0
Egress–AS	—	6	5
Metals technology–AS	—	5	1
NDI–AS	—	5	1
Structural repair–AS	—	12	1
Supervision	—	4	0
Total authorizations	216	44	40

We then need to perform new LCOM runs to identify the AS remainder for the backshops that only partially moved to the CRF. We cannot use the UMD in this case, because we have defined a new capability that supports only some fraction of the workload that was previously being performed by these shops. The LCOM analysis suggests that we need to move roughly 44 people from positions that were previously in the EMS or CMS into this AS (second data column of Table 3.3, "EMS/CMS Movement").

On the basis of our review of the OSD guidance and our discussions with ACC personnel, it became apparent that the ability to conduct split operations, wherein F-16 squadrons deploy some fraction of their PAA but also leave some of it at home station, is consistent with both programming guidance and recent experience. We performed additional LCOM runs to identify what the additional manpower requirement would be if we were to staff all CC squadrons in accordance with a split-operations capability. For the representative 24-PAA CC squadron mentioned above, if we assume a split-operations construct wherein 12 aircraft are deployed and 12 aircraft are operating at home station, this would generate a requirement for about 40 additional positions not currently in the UMDs (third data column of Table 3.3, "Split-Operations Plus-Up").[8]

Applying that logic across the F-16 fleet, we obtained the total authorizations presented in Table 3.4. The first row shows the existing aircraft maintenance squadron UMD, which we did not modify.[9] The

---

[8] In discussions with ACC/A4F16 personnel, we were informed that this split-operations manpower was added to the CC ACC F-16 UMDs (it was referred to as "AEF [Air and Space Expeditionary Force] manpower") in July 2004, although it was later removed from the manpower authorizations as part of the PBD 720 manpower reductions in the first quarter of FY 2007.

[9] Because we included the entire AMXS manpower in the AS requirements of Table 3.4, the weapon maintenance manpower supporting aircraft phase at the CRF constitutes a purely additive manpower requirement within this analysis. It would be possible to reduce the AS weapon maintenance manpower in light of the removal of phase-related workloads from the AS responsibilities; however, we do not consider those reductions in this analysis, and thus we generate an overestimate of the requirement for this work center.

Table 3.4
Manpower Requirements for F-16 Aircraft Squadron Operations

| Operation | AD | Manpower Authorization | | | | Total |
| | | ANG | | AFRC | | |
		Part-Time	Full-Time	Part-Time	Full-Time	
AMXS FY 2008 UMD	6,147	2,628	1,674	413	281	11,143
Moved from EMS/CMS	958	512	326	52	36	1,884
Split-operations plus-up	764	643	409	48	32	1,896
Proposed new AS	7,869	3,783	2,409	513	349	14,923

second row shows those positions that were moved from their previous assignment in the EMS or CMS to provide their new split-shop support to the AS. The third row presents the additional manpower requirement if all CC squadrons had their AS sized in accordance with a split-operations capability, applying that to not only the AD but also the ANG and AFRC personnel. For RC squadrons, we used a split-operations construct, in which a notional 18-PAA squadron would be staffed to support six deployed aircraft and 12 aircraft at home station, consistent with ANG "rainbowing" practice. Adding a split-operations capability across the TF generates a requirement for approximately 1,900 additional maintenance manpower positions that do not currently exist in Air Force UMDs. In essence, this approach has reassigned roughly 1,900 positions that were previously in the backshops into this AS and has created an additional requirement for 1,900 split-operations positions. In the following section, we present alternatives for paying for those positions, using savings from centralization of the component repair and phase-related workloads.

F-16 Repair Network Design Options

We now turn our attention to evaluation of the F-16 repair network of CRFs. The CRF workload and, by extension, its manpower requirements are primarily a function of the aggregate flying hours supported. As part of our LCOM analysis (discussed in the previous section), we identified the fraction of the total CRF workload that is associated with phase-related tasks. For the LCOM scenario that we simulated, approximately 60 percent of the CRF work centers' aggregate maintenance man-hours was associated with phase-related workloads; the remainder was associated with component repair. To identify the effect of networked CRF maintenance on logistics system performance, we need to understand aircraft phase production times (for phase-related workloads) and component pipeline times and requirements (for component repair workloads).

Figure 3.2 presents an analysis of the "fly-to-fly" times associated with F-16 phase inspections, differentiated by major command (MAJCOM) or component, collected over the interval October 2004–October 2007. The x-axis presents the number of days between the last aircraft sortie immediately preceding a phase inspection and the first aircraft sortie immediately following the phase, obtained from the

Figure 3.2
Phase Fly-to-Fly Times

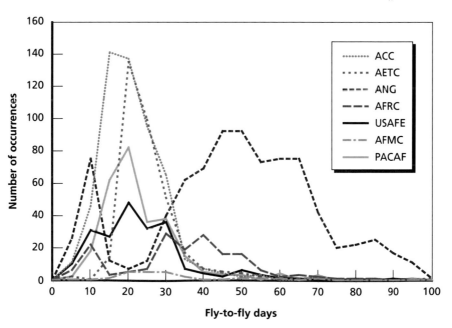

REMIS[10] data system. The y-axis presents the number of occurrences for each fly-to-fly interval, by MAJCOM or component, over this three-year interval.

Phase duration affects system performance in that aircraft do not fly missions for the entire fly-to-fly interval while they are in the phase dock. Figure 3.2 shows that AD aircraft accrue roughly 20 fly-to-fly days per F-16 phase, while ANG and AFRC aircraft accrue approximately 50 and 30 days, respectively. Given the PAA in each of the AD, ANG, and AFRC and the implied number of annual phases generated by the programmed flying-hour schedules, we can compute the average number of unavailable aircraft due to phase at any point in time, as shown in Table 3.5. This suggests that, in aggregate, the ANG would expect to have 50.5 aircraft unavailable due to phase at any point in time. Since only 23 ANG units are equipped with the F-16, this implies that an average unit would expect to have two of its aircraft unavailable due to phase maintenance at any point in time.

These implications are somewhat surprising, in view of the fact that a typical F-16 squadron has one phase dock in its maintenance facilities.

Table 3.5
Number of Unavailable Aircraft, Based on REMIS Data

MAJCOM/ Component	PAA	Phases/Year	Fly-to-Fly Days	Unavailable Aircraft
AD	603	489	20	26.8
ANG	423	369	50	50.5
AFRC	48	53	30	4.4
Total	1,074	911	—	81.7

---

[10] The Air Force's REMIS is a single, primary Air Force system for collecting and processing equipment maintenance data, which are used to provide information on reliability and maintainability, trend analysis, failure prediction, and weapon system availability. To support this analysis, we drew data from REMIS for the October 1, 2004–October 1, 2007 period. For further details regarding REMIS data analysis of both phase fly-to-fly times and phase throughput times, see Appendix C.

As an alternative, assume that a typical squadron would expect to have one aircraft in phase maintenance at any point in time, consistent with the single phase dock per F-16 squadron. Table 3.6 presents the fly-to-fly days implied by the assumption that a squadron would have one aircraft in phase maintenance at any point in time. While this time is similar to the mean fly-to-fly time presented in Figure 3.2 for ACC (18 days, compared with 20 days), the time is very different for ANG (23 days versus 50 days). These discrepancies may warrant further analysis, if one wished to understand the causes (including potential data errors) and implications of these long intervals.

The ANG and AFRC fly-to-fly times are much longer than the times for AD units. However, these intervals do not necessarily indicate the amount of time spent performing a phase, since within any fly-to-fly interval, there would likely be some amount of time during which no aircraft maintenance takes place. The fact that most ANG and AFRC maintenance units operate in a one-shift-per-day, five-days-per-week work environment, while AD units typically have two maintenance shifts per day, likely explains much of the difference between fly-to-fly intervals.

To identify the amount of time an aircraft would be unavailable because of phase in a CRF with an undetermined work schedule, we need to measure the phase throughput time, excluding all hours when no maintenance occurs. Figure 3.3 presents such throughput times for the F-16 fleet during 2004–2007, differentiated by aircraft block number.[11] We obtained the data presented here by the following process:

Table 3.6
Number of Fly-to-Fly Days, Assuming One F-16 in Phase Per Squadron

MAJCOM/ Component	PAA	Monthly Flying Hours/PAA	Squadron Phases/Year	Implied Fly-to-Fly Days
ANG	18	22	15.8	23.0
ACC	24	28	20.2	18.1

---

[11] At the end of FY 2008, there were 173 Block 25 PAA, 303 Block 30 PAA, 183 Block 40 PAA, 157 Block 42 PAA, 178 Block 50 PAA, and 80 PAA in other blocks.

Figure 3.3
Phase Throughput Times

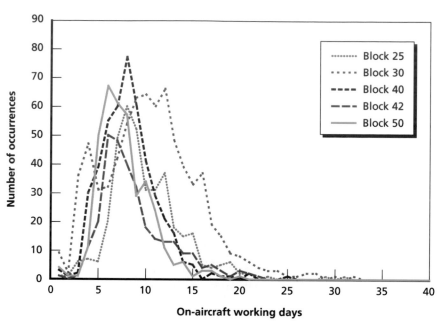

(1) identify by Work Unit Code (WUC) every incident of an F-16 phase; (2) identify the fly-to-fly interval associated with this phase inspection; (3) for every hour within this interval, identify whether any maintenance occurred (whether or not this maintenance is indicated with a "phase" maintenance-type code); (4) sum all hours in which any maintenance occurred. The x-axis shows the number of phase throughput days. The y-axis shows the number of occurrences for each phase throughput, by block number, over the three-year interval. When we remove those hours during which no maintenance occurred, the mean phase throughput times range from seven to ten days.

Because our analysis uses LCOM as the primary source for determining CRF manpower requirements, we examined the extent to which data reported in standard Air Force data systems, such as REMIS, were consistent with the assumptions inherent in the LCOM data. The accuracy of LCOM data was a potential cause for concern, because when we conducted the continental United States (CONUS) CIRF analysis in support of RE21 CIRF initiatives (McGarvey et al., 2008), we observed large discrepancies between the data as reported in LCOM and what was recorded in the standard data systems. As an example, that analysis suggested that the LCOM estimate of mean F110-100 JEIM repair flow time was 176 hours, while the Comprehensive Engine Management System indicated that the field was spending an average of 279 hours on such engine repair (even when overnight and weekend periods when no maintenance was occurring were eliminated). We do not see such large discrepancies here. The LCOM Block 40 estimate for a phase duration is about 7.5 days if round-the-clock operations are conducted. That duration is broadly consistent with the REMIS Block 40 performance presented in Figure 3.3, which has a mean throughput time of 7.7 days. Thus, we are confident that the LCOM estimates are fairly accurate in this case.

We next assess the effect of F-16 phase maintenance on aircraft availability, depending upon the work schedule that is assumed. Table 3.7 presents the daily worldwide average of the number of unavailable F-16 aircraft due to phase under four work schedules (assuming eight-hour shifts and an eight-day flowtime in each case) in support of programmed flying-hour schedules. Because the CRF option involves potentially removing phase maintenance from the operating unit and doing this work at a network facility, we also need to take into account the time an aircraft would be unavailable to perform missions because it is in transit to and from the CRF. If we assume an average of one day in transit, each way, to the CRF,

Table 3.7
Effect of Phase Management on Fleet Aircraft Availability

8-Hour Shifts/Day	Work Days/ Week	Calendar Days/ Phase	Transit Days to CRF	Worldwide Daily Average Unavailable Aircraft
3	7	8.0	2	24.7
3	5	11.2	2.8	34.6
2	7	12.0	2	34.6
2	5	16.8	2.8	48.5

aircraft are unavailable for an additional total of two days (or 2.8 days, when the CRF operates only five days per week). This suggests that, even using a two-shifts-per-day, five-days-per-week work schedule, regionalized CRF phase maintenance would not be expected to have a significant negative effect on aircraft availability, either when compared with the worldwide unavailability value computed in Table 3.5 (81.7 aircraft) or when using the rule of one unavailable aircraft per F-16 squadron, as was done in Table 3.6 (53 aircraft, the sum of 39 CC F-16 squadrons and 14 non-CC "24-PAA squadron equivalents").

In the remainder of this analysis, we assume 24-hours-per-day, seven-days-per-week CRF operations. On deployment at a contingency CRF, we assume standard factors of two shifts per day, 60 hours per week per person. We assume three shifts per day, 40 hours per week per person at a CRF supporting home-station operations.[12]

Removing CRF workload from the aircraft's operating location and assigning it to a network facility requires that failed aircraft components be transported between the aircraft operating location and the CRF. An inventory of spare components would also be required to support the delay time interval between a component's failure at the aircraft operating location and the receipt of a serviceable replacement from the CRF. For this analysis, we focused on the set of aircraft components appearing in both (1) the current readiness spares package (RSP) for any Air Force F-16 unit and (2) the RAND March 2006 capture from the D200 Requirements Data Bank (RDB) data system. Across all F-16 series and block numbers, this intersection comprises a set of 350 unique National Item Identification Numbers (NIINs).[13] For more details on this pipeline analysis, see Appendix F.

---

[12] Were a less-demanding work schedule assumed at home station, the home-station aircraft availability would decrease consistent with the values presented in Table 3.7, although the manpower requirements would not be expected to change significantly.

[13] There are 6,372 unique NIINs across all Air Force F-16 units' RSPs. However, 89 percent of these NIINs have an expendability, recoverability, reparability category (ERRC) code of XB3, indicating that they are "expendable not reparable"; such items would generally be disposed of upon failure and would thus not enter into any repair pipeline. Four percent of the RSP NIINs have ERRC code XF3 (authorized for repair at the field level and generally

The effect on transportation and inventory associated with the removal of CRF work centers from the aircraft operating location is limited to that fraction of component failures that were previously repaired at the on-site backshops.[14] We computed an estimate of this effect by multiplying the expected number of component failures by one minus the base-not-reparable-this-station (BNRTS) rate for each component.

Assuming a notional home-station monthly flying schedule of 27 flying hours per PAA, for the set of NIINs under consideration, and counting only those failures that would currently be repaired on site, we would expect to observe a daily fleetwide average of 74.9 component failures; however, because 35.4 of these failures would be associated with the shops that have been excluded from this CRF analysis (i.e., JEIM, electronic warfare, LANTIRN, and avionics backshops), only 39.5 of them would relevant to this analysis. To estimate the transport costs associated with the use of CRFs in support of home-station operations (including operations for permanently assigned Pacific Air Forces [PACAF] and U.S. Air Forces in Europe [USAFE] forces), we assumed that all failed components would be shipped using Federal Express (FedEx) Small Package Express two-day rates for U.S. domestic shipments.[15] Note that we are not endorsing FedEx here, merely utilizing their cost structure in an attempt to estimate the shipping costs, since FedEx is commonly used for shipping such parts. Focusing solely on those workloads that were formerly performed within

---

condemned when the field level cannot return them to serviceable condition), and 7 percent have XD2 (authorized for repair at the depot). Our D200 data set for the F-16 contains 1,870 unique NIINs, of which 1,830 have ERRC code XD2, four have ERRC code XF3, and 36 have ERRC code XB3. For the 350 NIINs included in this analysis, 341 have ERRC code XD2, two have ERRC code XF3, and seven have ERRC code XB3. The lack of XF3 components might understate the effect of backshop centralization, since it would generate a new pipeline requirement for such items.

[14] The assumption that the transportation of items between the operating unit and the depot would not be affected under this option would be valid as long as the remaining organizational level maintenance would be able to identify those items that require depot maintenance.

[15] FedEx rates are available on the U.S. General Services Administration Federal Supply Service Web site.

the backshops for the limited set of work centers under consideration, the expected annual fleetwide transportation cost in support of home-station operations is approximately $700,000.

An inventory requirement can be similarly computed. As in the transportation computations, an additive inventory requirement would be necessary to support the new transportation segments introduced by the CRF. The assumed two-day transport time in each direction generates a requirement for four days' worth of pipeline inventory. If we assume that a separate inventory requirement is computed to support each of the permanently assigned USAFE, PACAF, and CONUS F-16 fleets, operating at a notional flying schedule of 27 flying hours per month, a total one-time inventory investment of $4.8 million would be required to support home-station operations.

Because this inventory requirement is a one-time additional investment, the cost could be amortized across the expected duration of F-16 CRF operations. Considering an amortization interval as short as five years produces an annualized inventory requirement cost of less than $1 million. Further, a transportation pipeline and inventory requirement would not necessarily be created for every unit, e.g., if the CRF was at an existing F-16 operating location. Thus, because the cost associated with CRF component repair transportation and inventory is relatively small, we will not include it in the remainder of this analysis, focusing instead on the other, much larger system costs.

Removing workloads from the aircraft's operating location and assigning them to a CRF network generates a number of other costs, including those of shuttling aircraft (for phase inspections) and constructing new centralized facilities; all these costs will be discussed in more detail later in this chapter. The primary motivation for centralization, from a cost-benefit standpoint,[16] is the potential for reductions in maintenance manpower due to economy-of-scale effects.

Figure 3.4 presents our LCOM-derived estimate of these scale-economy effects for F-16 CRF maintenance; these results are particularly important and will motivate much of the remaining analysis. The

---

[16] Less-quantitative benefits, such as increased standardization of work practices, could also be realized via centralization.

Figure 3.4
F-16 CRF Manning Requirements, Home Station

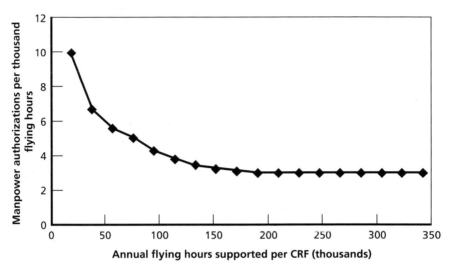

RAND *MG872-3.4*

x-axis shows the annual flying hours supported at a CRF. The y-axis shows the CRF maintenance manpower authorizations required per 1,000 flying hours. The LCOM analysis that was performed to identify the points lying on this curve is detailed in Appendix B. The left endpoint of the curve demonstrates that for a relatively small facility supporting a relatively small amount of flying, approximately ten manpower authorizations are required per 1,000 annual flying hours. The right extreme of the curve shows that a CRF supporting a much larger workload volume is able to achieve the same levels of performance (in terms of phase throughput times and simulated not-mission-capable-for-supply [NMCS] rates) with significantly less manpower. The reduction per thousand hours is from ten to about three. Note, however, that beyond 200,000 flying hours, the slope of the curve flattens, demonstrating no additional marginal reduction in manpower for facilities supporting more flying hours. These very large manpower savings require further explanation.

The strong scale economies arise because larger maintenance operations are able to achieve higher utilization of personnel. Smaller, decentralized maintenance operations have relatively low manpower utilization due to (1) minimum-crew-size effects, which are driven by the work center task that requires the largest crew size, even if most work center tasks require a smaller crew, and (2) "insurance" effects that take into account the fact that the maintenance organization needs the capacity to accommodate random spikes in demand without too great an adverse effect on flying operations.

Conversely, centralized maintenance operations with high workload volumes are able to achieve higher manpower utilization for the following two reasons: (1) the pooling of workloads reduces minimum-crew-size effects, and (2) the impact of variations in demand is dampened by the pooling of demands, reducing "insurance" requirements. To illustrate the first reason, consider the egress work center. LCOM analysis suggests that an egress shop supporting 24 PAA flying a sustained wartime schedule requires four manpower positions per shift, because of the minimum crew size of four. The LCOM analysis suggests that such a shop would achieve only a 14 percent direct labor utilization rate. Now consider instead a centralized egress shop supporting 20 such squadrons. Because the total manpower requirement for such a shop will be larger than the minimum crew size, the shop's manpower can be sized in accordance with the workload to be supported, allowing higher manpower utilization. It also becomes easier to assign maintenance workers to jobs, because many maintenance tasks are available to be performed at any point in time.

Pooling demands also decreases the effects of variations in demand. Because of the random fluctuations associated with both the failure process and the duration of maintenance activities, if a manpower utilization close to 100 percent were allowed, significant queues of unavailable components and aircraft would be expected. Smaller maintenance operations must maintain quite low manpower utilization (less than 20 percent for backshops supporting the 24-PAA squadron mentioned above), independent of minimum-crew-size effects, to ensure that adequate capacity is available to accommodate spikes in demand or repair durations. Because of these demand-pooling effects, backshops that are supporting ten such squadrons can meet the same level of performance (measured in terms of sortie success rate, total not-mission-capable for supply (TNMCS), and maintenance production rate) at a maximum manpower utilization of 45 percent.

Those two effects enable us to increase manpower utilization, as shown in Figure 3.5. The x-axis is the same as that in Figure 3.4, annual flying hours supported per CRF. The y-axis presents the average percentage direct utilization for maintenance manpower. At the left end of the curve, for decentralized, smaller operations (consistent with the annual flying hours for one deployed 24-PAA squadron), the average direct utilization for manpower is approximately 18 percent. Moving to the right endpoint of the curve, the more highly centralized location supporting a higher volume of flying is able to achieve significantly higher manpower utilization, on the order of 46 percent, due to the two effects mentioned above. The roughly 3:1 ratio presented here is consistent with the 3:1 ratio presented in Figure 3.4.[17]

Figure 3.5
Effect of CRF Workload on Manpower Utilization

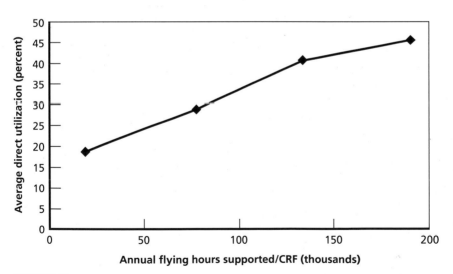

RAND MG872-3.5

_____

[17] These ratios are not identical due to the effects of indirect labor, along with the supervisory and support manpower positions (which are included in Figure 3.4 but not in Figure 3.5, since the concept of "direct utilization" is unclear for non–direct-labor personnel).

Given the dramatic reductions in maintenance manpower that our LCOM analysis suggests can be realized by centralization, to the extent that maintenance manpower costs dominate the other costs under consideration (transportation, construction, equipment), highly centralized solutions would be preferable.

Flexible Network to Support Contingency Operations

Having examined the maintenance manpower requirements associated with CRF operations, we now turn our attention to the design of the CRF network, focusing first on the flexible network necessary to support contingency operations. In this analysis, we begin with the requirements of the contingency-deployed fleet. Once these requirements are identified, we then "force" the fixed network that supports home-station operations to accommodate the contingency requirements, rather than beginning with an optimization for home station and then forcing the contingency support to fit within what is best for home-station operations.

To develop optimization approaches to evaluate global CRF network designs, we first identified the contingency support requirements for deployed units. We then developed an optimization model that evaluates designs for the home-station, i.e., "fixed," network, considering a set of resource trade-offs. We next developed total manpower requirements by integrating the contingency and fixed networks, while considering other policy decisions, such as the rotational burden on personnel and the AD/RC mix.

Six factors drive the contingency support network:

- size of the deployed fleet (PAA)
- intensity of OPTEMPO (utilization rate [UTE] and average sortie duration [ASD], and duration of deployment)
- in-theater dispersion (number of operating locations and distance between them)
- distance of deployed operating locations from fixed maintenance network sites
- available infrastructure (maintenance shelters).

The product of the PAA, UTE, ASD, and deployment duration determines the aggregate flying hours to be supported, which in turn defines the requirement for forward-deployed phase and component repair workloads. The dispersion of in-theater operating locations and the distance between them and the fixed maintenance network facilities (whether at CONUS, PACAF, or USAFE) determines the geographic dimension associated with aircraft shuttling and component transport. Finally, an understanding of the available maintenance infrastructure in the theater can influence both the decision to centralize in-theater maintenance and the decision of where to locate a CRF.

In this analysis, we consider three support options for contingency operations. In the first, each FOL operates without network support. An alternative way to view this option is that it has a mini-CRF deploy to each FOL to provide its support without relying on connectivity to the network. The second option is to establish a contingency CRF (CCRF) in the theater to support a number of FOLs. The third option is to retrograde aircraft and components from the FOLs to a fixed network site for CRF support.

Each of these options has associated risks in a contingency environment. Deploying CRF maintenance capabilities to each FOL places a large logistics footprint in forward-deployed locations, exposing CRF personnel and equipment to risks, such as potential attacks on the FOL, and increasing the forward base population that needs housing, facilities, and security-force protection. It also increases the requirements for deployment movements and lengthens deployment time lines. However, options that rely on off-site CRF support via connectivity to the repair network are at risk of potential disruptions to the transportation network that links FOLs to CRFs. Were a CCRF established in the theater, the loss of access to such a site, perhaps through enemy attack or natural disaster, would be more catastrophic than the loss of maintenance at a single FOL in a non–repair-network environment. Although these risks vary across maintenance support concepts, we can still compute and directly contrast the manpower and transportation requirements associated with each option.

Figure 3.6 presents a screen capture from a decision support tool that we developed to help identify, for a given deployment, the relative

Figure 3.6
Contingency CRF Decision Support Tool

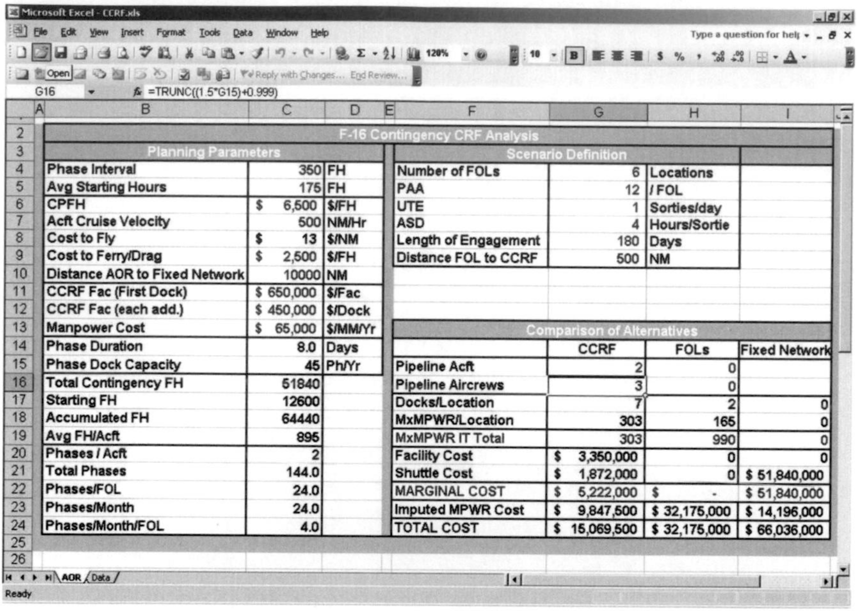

Planning Parameters			Scenario Definition		
F-16 Contingency CRF Analysis					
Phase Interval	350	FH	Number of FOLs	6	Locations
Avg Starting Hours	175	FH	PAA	12	/FOL
CPFH	$ 6,500	$/FH	UTE	1	Sorties/day
Acft Cruise Velocity	500	NM/Hr	ASD	4	Hours/Sortie
Cost to Fly	$ 13	$/NM	Length of Engagement	180	Days
Cost to Ferry/Drag	$ 2,500	$/FH	Distance FOL to CCRF	500	NM
Distance AOR to Fixed Network	10000	NM			
CCRF Fac (First Dock)	$ 650,000	$/Fac			
CCRF Fac (each add.)	$ 450,000	$/Dock			
Manpower Cost	$ 65,000	$/MM/Yr	Comparison of Alternatives		
Phase Duration	8.0	Days		CCRF	FOLs	Fixed Network
Phase Dock Capacity	45	Ph/Yr	Pipeline Acft	2	0	
Total Contingency FH	51840		Pipeline Aircrews	3	0	
Starting FH	12600		Docks/Location	7	2	0
Accumulated FH	64440		MxMPWR/Location	303	165	0
Avg FH/Acft	895		MxMPWR IT Total	303	990	0
Phases / Acft	2		Facility Cost	$ 3,350,000	0	0
Total Phases	144.0		Shuttle Cost	$ 1,872,000	0	$ 51,840,000
Phases/FOL	24.0		MARGINAL COST	$ 5,222,000	$ -	$ 51,840,000
Phases/Month	24.0		Imputed MPWR Cost	$ 9,847,500	$ 32,175,000	$ 14,196,000
Phases/Month/FOL	4.0		TOTAL COST	$ 15,069,500	$ 32,175,000	$ 66,036,000

RAND *MG872-3.6*

costs and benefits associated with performing CRF maintenance under the options mentioned above. We developed this tool in an attempt to identify the "breakpoints" above or below which it makes sense to establish a forward-deployed phase maintenance capability. The blue numbers in the spreadsheet indicate the input factors—information on both the scenario to be supported and some cost and manpower requirements. The tool takes these inputs and generates the output presented in the "Comparison of Alternatives" in the lower right of the figure. The alternatives are the three presented earlier: CCRF (establishment of a CRF in the theater to support a number of FOLs), FOLs (each of which provides its own support by deployment of CRF personnel without connectivity to the network), and a fixed network (reachback to the fixed maintenance network).

To evaluate the relative performance of these options, a number of trade-offs need to be considered. The primary trade-off occurs between the increased transportation requirement associated with off-site CRF support and the potential reductions for forward-deployed maintenance manpower under the CCRF (or fixed-network) options. An operational effect is associated with the aircraft pipeline requirement: If aircraft receive phase maintenance at an off-site CRF, those requiring phase must fly between the FOL and the CRF, making them unavailable for their operational mission during that interval. Given the scenario characteristics presented in Figure 3.6, for this notional deployment of 12 PAA to each of six FOLs, there would be a pipeline requirement to deploy two additional aircraft (beyond the 72 deployed to the FOLs) and three additional aircrews to support that aircraft pipeline. This pipeline requirement can be contrasted with the forward-deployed requirement for maintenance manpower (presented in the row labeled "MxMPWR IT Total"), allowing the analyst to identify what these two pipeline aircraft "buy" in terms of reduced forward-deployed maintenance manpower. The focus is on aircraft and aircrews that are required to be in the theater. We assume that the fixed network would swap an aircraft and crew into and out of the theater. A pipeline would exist, but not in the theater. Therefore, in the comparison, this cell is left blank because it is not entirely applicable. While the CCRF option for this scenario requires approximately 300 people, the FOL-unique support option requires nearly 1,000 forward-deployed maintenance manpower positions. Beyond the simple increase of 700 maintenance personnel, the FOL-unique option incurs additional costs (not presented in the decision support tool) associated with providing base support, security, etc., for these additional maintainers.

The remaining rows in the comparison table provide additional information allowing for the contrast of some other cost implications associated with these maintenance options. If the Air Force sees value in this decision support tool, the tool could be refined to capture any additional considerations that are deemed germane.

Identifying CRF Manpower Requirements Across a Range of Scenarios

A maintenance manpower force-sizing analysis needs to address the capability of the maintenance force to support a variety of potential deployment requirements. Table 3.8 presents the forward-deployed

Table 3.8
Manpower Requirements Across a Range of Scenarios

Type of Network	Manpower Authorization					
	0% deployed	SSSP, 10% deployed	1 AEF, 20% deployed	2 AEF, 40% deployed	MCO, 80% deployed	100% deployed
1 CCRF	—	285	421	803	1,563	1,981
2 CCRFs	—	416	570	842	1,606	1,986
3 CCRFs	—	495	753	909	1,608	2,067
4 CCRFs	—	660	832	1,140	1,684	1,992
5 CCRFs	—	825	1,040	1,340	1,780	2,105

NOTE: We use the terms *SSSP* and *MCO* in a notional, but representative, sense.

maintenance manpower requirements evaluated across a range of engagement scenarios. The top row presents the fraction of the CC aircraft that are deployed for a set of scenarios ranging from 0 percent to 100 percent of the CC fleet. The leftmost column shows the number of forward CCRFs established in support of each scenario. The values in the rows indicate the number of forward-deployed maintenance manpower positions necessary to support a given deployment level with a given number of CCRFs (assuming that the deployed aircraft are allocated equally across all CCRFs). For example, a scenario in which 20 percent of the CC F-16s are deployed and are supported by four CCRFs (with 5 percent of the aircraft being supported at each CCRF) requires a total of 832 maintenance manpower positions.

For any scenario in Table 3.8, the forward-deployed manpower requirement can be translated into a total maintenance force-sizing construct. Rather than suggesting that the Air Force build its maintenance manpower capability to any specific level in this table of alternatives, we present in the remainder of this section an exemplary force-sizing analysis built around a notional desired capability level. The analytic process we use could be applied to any other desired capability level. We note, however, that the OSD guidance discussed in Chapter One directs the services to plan and program for a future in which some fraction of the force is continuously forward-deployed.

For the purposes of illustration, assume that the Air Force desires the capability to maintain a continuous deployment of 10 percent of the CC F-16s into two theaters, with a single CCRF in each theater.[18] The shaded cell in Table 3.8 represents this scenario, which requires 416 maintenance manpower positions.

A number of policy alternatives need to be identified to support this requirement for an indefinite duration. We first need to identify the fraction of this workload that should be assigned to the AD. In this illustrative analysis, suppose that 90 percent of that forward-deployed workload is to be supported by the AD force. We next need to identify an acceptable rotational burden to place upon AD maintenance personnel. Assume that an AEF-like construct with a 4:1 dwell-to-deployment ratio (that is, for every day an individual is deployed, he spends four days at home station) is deemed supportable. That implies that maintaining 374 forward-deployed maintenance manpower positions at a 4:1 ratio requires four times that number of manpower positions in the fixed network, suggesting that the fixed network needs a minimum of 1,496 maintenance manpower positions. With the earlier assumption that 90 percent of the forward-deployed workload is to be supported by AD personnel, the remaining 10 percent of the requirement must be supported by RC personnel. We then need to identify an acceptable volunteerism rate for the RC personnel. Let us assume that RC drill personnel are available for a 9:1 dwell-to-deploy ratio, that is, one-half the deployment burden associated with the AD. Because the RC is responsible for supporting the remaining 42 forward-deployed positions, this generates a requirement for 420 RC drill positions. However, to train these RC drill personnel for their maintenance responsibilities, there is an additional requirement for full-time RC technicians. Assuming a 1:8 trainer-trainee ratio produces a requirement for 53 full-time RC technicians. Summing the 374 forward-deployed AD personnel, 1,496 home-station AD maintainers, 420 RC drill positions, and 53 full-time RC technicians produces a lower bound of 2,343 for the contingency scenario's maintenance manpower requirement. This is a lower bound because it is uncertain whether the 1,496 home-station

---

[18] We have not assumed that the deployment requirement varies across future years, as was presented in Figure 2.1. Rather, we have identified a capability level that can sustain a given level of deployment activity across the entire steady-state planning horizon.

positions will suffice to support the training flying schedule's maintenance requirement.

Fixed Network to Support Home-Station Operations

Having determined the maintenance manpower necessary to provide the desired capability level for continuously deployed forces, we next turn our attention to the fixed maintenance network that supports training and readiness missions at home station across all AD, AFRC, and ANG F-16 units. We developed an optimization model to identify the most cost-effective fixed maintenance network; further details of this integer linear programming (ILP) model are presented in Appendix D. For the fixed network, we focus on the performance of CRF alternatives with respect to their total system costs. We do not focus on operational metrics for the fixed network, because we assume that (1) all CRF alternatives identified will have sufficient maintenance manpower to accomplish all required home-station workloads without harming operational performance due to the buildup of large maintenance queues, and (2) the operational effect of flying home-station aircraft between their operating location and a phase CRF is unclear, since the aircraft are, by definition, "mission capable" during this sortie, and this sortie could potentially be used to accomplish training missions. Note, however, that we assume the shuttle cost between home station and the phase CRF is a 100 percent additive cost, which is equivalent to saying that no training missions are accomplished during this sortie. This assumption makes centralization less attractive and would tend to favor less-centralized networks, to the extent that these shuttling costs are relatively large. This optimization model considers the entire range of CRF network alternatives, from fully decentralized solutions that retain CRF maintenance capabilities at all sites to fully centralized alternatives that consolidate all CRF capabilities at one site, and identifies the alternative that minimizes the total cost.

The parameters that influence the design of the optimal fixed network are similar to those that were considered for the contingency network. Key inputs include the flying schedule to be supported (we used the FY 2008 programmed flying hours from AFTOC CAIG), the extent of labor-scale economies, the minimum manpower needed to staff CCRFs at an acceptable deployment burden, and some additional cost data associated with personnel, aircraft shuttling, and facilities.[19] We assumed a personnel cost of $65,000 per manpower authorization (U.S. Air Force, 2007a). To determine aircraft shuttling costs between operating locations and potential CRF sites, we used the average F-16 cost per flying hour (CPFH) of $6,500 from AFTOC CAIG, with flying hours between any two points determined by dividing the flying distance between them by an F-16 block-speed planning factor of 600 nmi/hr. The only facility cost included in this analysis was the cost associated with constructing new aircraft hangars at the CRF sites. We obtained an annualized cost of $100,000 per phase dock by taking the construction cost and amortizing it over 15 years.[20] This analysis assumes that existing hangar facilities could not be used at any CRF site and that new hangars to support phase operations would be required at any CRF. Some potential CRF sites currently have underutilized hangar space that could be used for CRF phase operations, but such sites were not allowed to take advantage of these potential cost avoidances in this analysis.[21]

We first considered the home-station support to F-16 units stationed in USAFE and PACAF (excluding the F-16 unit at Eielson AFB, which was to be supported by maintenance from CONUS). The optimization model indicates that these units can be served from a

---

[19] Equipment costs might also be associated with establishing CRFs. For shops that would be consolidated in their entirety at a CRF (e.g., pneudraulic), this would not be an issue—such equipment currently exists at each operating location, and this pool of equipment would be consolidated at the CRFs. However, for shops to be split between the AS and the CRF (e.g., NDI), additional equipment could be necessary, if, for example, each unit currently has only a single piece of some specialized item, in which case each AS would need to retain its own piece, requiring an additional procurement for each CRF's use. However, we did not investigate this matter and do not include any equipment costs in this analysis.

[20] Data for this cost were drawn from *Historical Air Force Construction Cost Handbook* (2004).

[21] During meetings with ACC/A4 staff, it was suggested that the facility costs might be somewhat understated in this analysis; the primary facility cost that was thought to be missing was the cost associated with constructing a "fuel barn" at a CRF. A fuel barn is an aircraft hangar that has special ventilation, electrostatic discharge protection, and breathing equipment for performing maintenance on aircraft fuel cells.

single CRF in each area of responsibility (AOR). Equally important, the analysis indicates that support to USAFE and PACAF units is relatively insensitive to the precise CRF location selected. That is, the USAFE CRF could be sited at either Aviano or Spangdahlem, with very little discernible effect on cost or operational performance. Similarly, in PACAF, the CRF could be located at any of the Japanese or Korean sites with very little effect on performance or cost. The analysis indicated that support to PACAF and USAFE required a total of 514 maintenance manpower positions.

However, the CONUS beddown is considerably more complex, as shown in Figure 3.7, because of the large number of locations, their relative dispersal, and the variance in the number of assigned aircraft between fairly large units (such as Luke AFB, which has 151 PAA) and rela-

Figure 3.7
FY 2008 F-16 CONUS Beddown

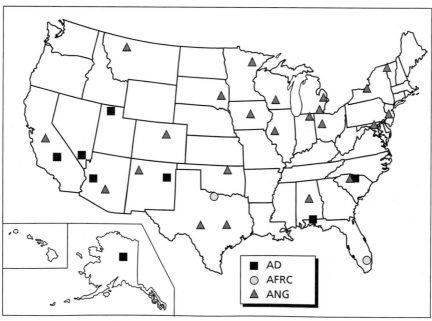

RAND *MG872-3.7*

tively small units (such as the many ANG units with 15 or 18 PAA). The AD unit at Eielson AFB was included in the CONUS analysis.

Our optimization model identifies the minimum-cost network as having two CONUS CRFs: one in the western United States (at Nellis AFB) and one in the eastern part of the country (at Springfield ANG, Ohio). However, many central locations would also be reasonable sites for a single CONUS CRF; the model identified Kirtland ANG, New Mexico, as the minimum-cost single CRF site. Table 3.9 presents the performance of a set of CONUS CRF alternatives. Cost and manpower details are shown for the minimum-cost network associated with different numbers of CONUS CRFs, as identified by the optimization model. Note that the facility costs are constant across all solutions; this occurs because it was assumed that all CRF facility requirements are purely additive costs applied on a "per aircraft inspection space" basis, and each of these solutions constructs the minimum number of spaces necessary to support the workloads assumed in this scenario. Note also that the total annual costs associated with the single-CRF and two-CRF networks are fairly comparable.

Figure 3.8 contrasts the performance of the best single-CRF and two-CRF networks. The single-CRF solution is able to achieve cost performance similar to that of the two-CRF network by essentially trading manpower for transport costs. A single CRF has a reduced manpower

Table 3.9
F-16 CONUS Fixed-Network Options

Item	Costs and Manpower Positions				
	1 CRF	2 CRFs	3 CRFs	4 CRFs	5 CRFs
Manpower costs ($M/year)	63.8	65.8	75.5	85.3	96.0
Shuttle costs ($M/year)	11.5	6.4	5.8	5.5	5.0
Facility costs ($M/year)	1.7	1.7	1.7	1.7	1.7
Total annual costs ($M/year)	77.0	73.9	83.0	92.5	102.7
Manpower positions	982	1,012	1,162	1,312	1,477

Figure 3.8
Costs of One and Two CONUS CRFs

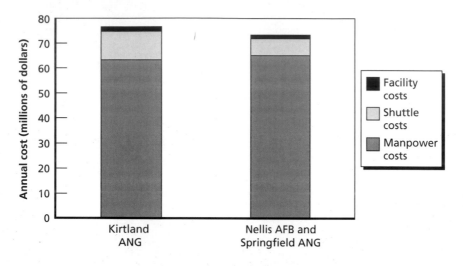

requirement because it is better able to achieve the high manpower utilization associated with increasingly centralized maintenance networks.[22] However, the single CRF generates a higher shuttle cost, because the average operating-location-to-CRF distance is greater for a single-CRF network than for a two-CRF network. Figure 3.8 demonstrates that CONUS F-16 CRF support is somewhat insensitive to the precise number of locations established. It is possible to establish either one or two locations with little effect on cost performance.

Our analysis also suggests relative insensitivity to the *precise locations* of CRFs. The two bars on the left of Figure 3.9 present the performance of the minimum-cost-solution networks from Figure 3.8. These

---

[22] In fact, the single-CRF manpower requirement has been inflated to 982 because of the forward-deployed-manpower rotational burden constraint that was imposed earlier: A requirement for a total of 1,496 full-time maintenance positions at home station, less the 514 positions assigned to PACAF and USAFE, implies that a minimum of 982 maintenance manpower positions must be assigned to the CONUS fixed network, even though the single-CRF solution actually requires only 939 manpower positions.

Figure 3.9
Alternative CRF Solutions

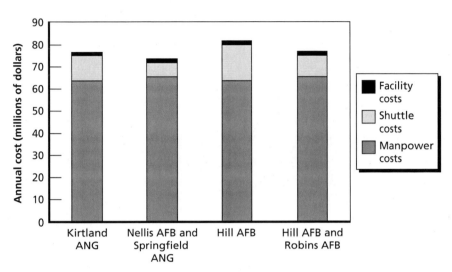

are contrasted with two alternative networks: a maintenance network with a single CRF established at Hill AFB and a two-CRF solution with CRFs established at Hill AFB and Robins AFB. Because the total cost is comparable across the solutions in Figure 3.9, we can conclude that F-16 CONUS CRF support is relatively insensitive to the precise locations selected for CRFs. This allows another set of considerations beyond the scope of this analysis to enter into the final CRF location decision. As an example, the establishment of a CRF at Hill AFB could also provide proximity to the F-16 SPO, or to depot personnel.

Once a fixed network is selected, the steady-state manning requirement is determined. Suppose that we wanted to implement the single-CONUS-CRF solution. When we add the 982 CONUS CRF positions to the 514 maintenance positions at the USAFE and PACAF CRFs, plus the contingency CRF personnel mentioned earlier, the total steady-state manning requirement sums to 2,343 positions.

While Table 3.8 presents the deployed manpower require-ment across a range of scenarios, it does not include the maintenance manpower requirement necessary to support home-station operations.

Table 3.10 updates these data by including the fixed network, showing the maintenance manpower requirement associated with the minimum-manpower fixed network (without including any deployment-burden constraints on the fixed network).

Recall that our analysis of steady-state requirements assumed that a scenario involving 10 percent deployment into two theaters defined the required maintenance capability, with an associated requirement of 416 forward-deployed positions. It is worthwhile to consider the performance of the CRF maintenance network against a range of potential futures. This capability level is also able to support the less stressing scenarios in Table 3.10, i.e., 0 percent deployment or 10 percent deployment in one theater, under the AD/RC mix and deployment-burden policies selected earlier. This maintenance capability could also support an AEF-like deployment of 20 percent of the forces into one theater (with its requirement for 421 continuously sustained forward-deployed positions). The 2,343 total maintenance manpower personnel computed above can also support a 40 percent deployment into one theater, although at the cost of a deployment burden double the policy design objective.

However, it is not sufficient to consider only the steady-state requirements. A maintenance manpower force-sizing analysis also needs to identify what is required to support surge MCO demands.

Table 3.10
Alternative CRF Manpower Requirements

	Manpower Authorization					
Type of Network	0 % deployed	SSSP, 10% deployed	1 AEF, 20% deployed	2 AEF, 40% deployed	MCO, 80% deployed	100% deployed
Fixed	1,512	1,453	1,395	1,210	926	441
1 CCRF	—	285	421	803	1,563	1,981
2 CCRFs	—	416	570	842	1,606	1,986
3 CCRFs	—	495	753	909	1,608	2,067
4 CCRFs	—	660	832	1,140	1,684	1,992
5 CCRFs	—	825	1,040	1,340	1,780	2,105

In the same manner as we chose a requirement for 10 percent deployment into two theaters as an illustration for the steady state, suppose that we now identify an 80 percent deployment of the CC fleet into two AORs, each of which has its own in-theater CRF, as the MCO capability-level requirement. Referring to Table 3.10, this would generate a total requirement of roughly 1,600 deployed personnel and 900 home-station positions, for a total maintenance manpower level of 2,532. The steady-state analysis identified a requirement for 2,343 positions, so MCO considerations generate a requirement for 189 additional maintenance manpower positions. One alternative would be to staff these 189 positions using RC drill personnel. Note the assumption that rotational-burden considerations do not apply to an MCO surge-type environment.

We repeat, however, that the force-sizing analysis presented in this section is not intended to represent a recommended capability level (although it is broadly consistent with OSD planning and programming guidance). Rather, it is meant to illustrate how the manpower requirements identified in this analysis can be translated into a total maintenance force-sizing construct with the selection of a few additional policy choices, such as the fraction of the CRF workload assigned to the AD. This analytic process could be applied to any other capability level the Air Force deemed appropriate.

CRF Networks to Support Only AD and AFRC Forces

In addition to the TF analyses, the Air Force asked for an analysis in which the repair network would support only the AD and AFRC units. Whereas the TF analyses reallocated manpower among all units and the repair network, these additional analyses reallocate only AD/AFRC manpower within the network; ANG manpower authorizations are not modified. In these additional analyses, we assume that all AD and AFRC units receive home-station support from the fixed CRF network and deployment support from a CCRF. We further assume that ANG units receive no support from the repair network.[23]

The research methodology and modeling tools used for these additional analyses are identical to those used for the TF analyses. We again use LCOM and the RAND-developed optimization tool. The methodological sequence is also identical. We first establish a baseline for the current maintenance manpower and then determine the requirements for an AS that performs all sortie launch and recovery maintenance workloads. We then determine the resource requirements for the CCRFs and the fixed network, in support of similar steady-state and MCO scenarios as were presented earlier.

Baseline Maintenance Manpower

Table 3.11 shows the F-16 maintenance manpower, assuming the ANG authorizations are not modified. Table 3.11 is organized basically like

Table 3.11
AD and AFRC F-16 Maintenance Manpower Authorizations

		Maintenance Manpower Authorization					
		ANG		AFRC			
Operation	AD	Part-Time	Full-Time	Part-Time	Full-Time	Total	Total AD/AFRC
Group and MOS	1,954	598	631	82	98	3,363	2,134
AMXS	6,147	2,628	1,674	413	281	11,143	6,841
CMS and EMS							
Propulsion and avionics	1,516	480	693	68	106	2,863	1,690
AGE and munitions	2,539	936	363	176	79	4,093	2,794
Phase and related	3,143	1,481	1,233	196	168	6,221	3,507
Total	15,299	6,123	4,594	935	732	27,683	16,966

SOURCE: F-16 FY 2008 UMD.

---

[23] This assumption could present difficulties during deployment: If, for example, both AD and ANG F-16s deployed to a common location, there could be two F-16 backshop maintenance concepts of operation (CONOPs) at that site. An alternative would be to have ANG aircraft supported via a CCRF during deployment but not receiving support from the fixed CRF network while at home station. This assumption also poses potential complications because ANG units would receive backshop maintenance support under one CONOP while at home station and under a different one while deployed, which would require ANG personnel to deploy and work according to the AS and CRF constructs.

Table 3.1; however the columns for part- and full-time ANG spaces are grayed out, as is the "Total" column. This indicates that these ANG positions have been excluded from our analyses. The rightmost column shows the total maintenance manpower if only AD and AFRC positions are considered. Excluding the ANG reduces the total manpower authorization by nearly 11,000 positions, or nearly 40 percent of the total. As before, the shaded cells indicate that the focus of this analysis is on a reallocation of manpower between the flightline and phase-related backshop positions.

AS Requirements

The requirements for AS maintenance were computed on the basis of individual squadrons. Thus, it is straightforward to exclude ANG squadrons from the associated restructuring of manpower necessary to create an AS at each squadron and to add a split-operations capability at each CC squadron. Table 3.12 modifies the results presented in Table 3.4, indicating that establishing an AS maintenance capability at AD and AFRC F-16 squadrons requires a total of approximately 9,000 positions, reassigning roughly 1,000 positions that were previously in the backshops into the AS and creating a new requirement for approximately 850 split-operations positions that do not currently exist in maintenance UMDs. Because we have assumed that ANG manpower is not modified in this analysis, all ANG manpower columns are grayed out, and ANG flightline operations are assumed not to receive the additional split-operations manpower computed earlier. The relative manpower increase associated with split-operations is larger for the ANG (24 percent of AMXS) than for either the AD or the AFRC (12 percent of AMXS for each). This implies that an AD/AFRC-only CRF network will need a smaller reduction in backshop positions because of centralization to realize a constant maintenance manpower total with the current baseline.[24]

---

[24] This set of assumptions leads to the creation of squadrons of differentiated capabilities, with AD and AFRC squadrons staffed to support split-operations and ANG squadrons lacking this increased capability.

Table 3.12
Manpower Requirements for AD and AFRC F-16 Aircraft Squadron Operations

| | | Manpower Requirement | | | | | |
| | | ANG | | AFRC | | | |
Item	AD	Part-Time	Full-Time	Part-Time	Full-Time	Total	Total AD/AFRC
AMXS FY 2008 UMD	6,147	2,628	1,674	413	281	11,143	6,841
Move from EMS/CMS	958	512	326	52	36	1,884	1,046
Split-operations plus up	764	643	409	48	32	1,896	844
Proposed new AS	7,869	3,783	2,409	513	349	14,923	8,731

CCRF Requirements

Our analysis of F-16 TF requirements indicated that 416 deployed CCRF positions were necessary to support an illustrative steady-state scenario involving the deployment of 10 percent of CC F-16s into two regions. We considered a similar scenario for the AD/AFRC-only analyses. Assume that the steady-state scenario to be supported involves the deployment of 72 PAA (approximately 10 percent of the TF CC PAA) across two theaters. Further assume that the ANG will provide 25 percent of these aircraft, or 18 PAA. This constitutes roughly 5 percent of the total ANG CC PAA. The remainder of the 54 PAA is drawn from the AD/AFRC fleet, accounting for approximately 13 percent of the total AD/AFRC CC PAA. Recall our earlier assumption that the deployed ANG aircraft are supported by the ANG at the deployed operating location, outside of the CCRF network. In this case, the CCRF manpower required to support a deployment of 54 aircraft can be computed, as was done in the TF analysis. Such a computation indicates that 330 manpower positions would be required to staff these CCRFs.

Fixed-Network Requirement

We used the optimization model described above (and in Appendix D) to identify alternatives for the fixed CRF network, now supporting only AD and AFRC forces. As indicated in Figure 3.10, the CONUS

Figure 3.10
FY 2008 AD and AFRC F-16 CONUS Beddown

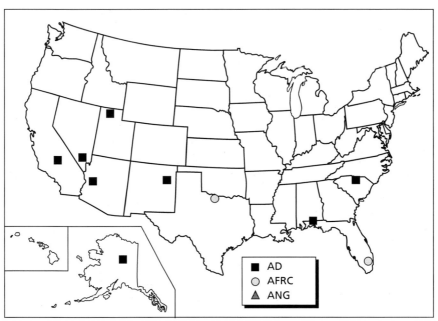

RAND *MG872-3.10*

F-16 network becomes much simpler with the exclusion of all ANG units, decreasing from 33 nodes to 10 nodes. Since all permanently stationed PACAF and USAFE F-16 units are AD, this additional analysis has no effect on the CRF options already described for these forces.

The optimization model identified a minimum-cost solution that establishes one CONUS CRF at Nellis AFB in addition to the single CRFs identified at PACAF and USAFE. The total manpower requirement associated with such a solution calls for 546 positions at the CONUS CRF and 514 positions summed across the PACAF and USAFE CRFs, for a total manpower requirement of 1,060. Recalling the steady-state scenario's CCRF deployment requirement of 330 positions, this manpower pool implies a dwell-to-deployment ratio of 3.2:1 (1,060/330). If the Air Force believed that such a ratio placed an undesirably high deployment burden on maintenance personnel, it could elect to increase the ratio by adding additional manpower to the fixed CRF network beyond the minimum requirement of 546; we assume that these additional positions would be assigned to CONUS instead of to PACAF or USAFE. Suppose that a dwell-to-deploy ratio of 4:1 is desired for all CRF personnel. Not only does this affect the manpower level, it also influences the design of the cost-optimal network.

A 4:1 ratio implies at least 1,320 manpower positions in the fixed network; subtracting the 514 positions associated with the permanent PACAF and USAFE CRFs results in a minimum manpower requirement of 806 CONUS CRF positions.[25] When such a manpower lower-bound constraint is enforced, the cost-optimal CONUS network has two CONUS CRFs, one at Nellis AFB and one at Shaw AFB. This occurs because the minimum-cost model is no longer able to bring maintenance manpower, the primary cost driver, to its minimal value. Because at least 806 positions are required in the CONUS network, the model is able to identify a two-CRF solution that can perform all the required maintenance workload, even at reduced scale economies (when contrasted with the single-CRF solution). Given that this two-CRF solution employs the minimum allowable manpower, the model then turns its attention to minimizing the transport costs, which necessarily decrease as the number of CRFs increases; were it possible to perform the entire CONUS workload at three locations using 806 manpower positions, the model would have selected such a solution. However, this is not possible because of the substantial reductions in scale economy associated with splitting the CONUS workload across three sites. Table 3.13 presents further details on the costs associated with these two solutions. Increasing the dwell-to-deploy ratio for CRF manpower from 3.2:1 to 4:1 raises annual personnel costs by $16.9 million but reduces annual shuttle costs by $2.9 million, for a net annual increase of $14 million and an associated increase of 260 manpower positions.

---

[25] The deployment burden on ANG personnel is not addressed here, but the scenario definition implies that the ANG must be able to support the sustained deployment of 18 PAA via its own backshop maintenance manpower.

Table 3.13
F-16 AD/AFRC CONUS Fixed-Network Options

Item	Costs and Manpower Positions	
	3.2:1 Dwell-to-Deploy Ratio, 1 CRF (Nellis AFB)	4.0:1 Dwell-to-Deploy Ratio, 2 CRFs (Nellis AFB and Shaw AFB)
Manpower costs ($M/year)	35.5	52.4
Shuttle costs ($M/year)	5.4	2.5
Facility costs ($M/year)	0.9	0.9
Total annual costs ($M/year)	41.8	55.8
Manpower positions	546	806

As was observed in the TF analysis, the AD/AFRC-only CRF network exhibits relative insensitivity to the precise locations for CRFs. Suppose that, as before, considerations beyond the scope of this analysis suggested that it might be desirable to establish an F-16 CRF at Hill AFB. This would have a minimal effect on the total costs. For either solution presented in Table 3.13, the manpower requirement would not vary with changes in CRF location, and neither would facility costs, since it was assumed that all CRF facility requirements are purely additive costs. The only cost that would vary with CRF location is the shuttle cost. For the single-CRF solution, locating the CRF at Hill AFB instead of Nellis AFB would increase the annual shuttle cost by $3.6 million over the total annual cost of $41.8 million for the minimum-cost Nellis AFB solution. Similarly, if a 4:1 dwell-to-deploy ratio is enforced and if a CRF at Hill AFB is mandated, the optimization model identifies a network with CRFs at Hill AFB and Luke AFB, with an associated annual shuttle cost increase of $1.7 million over the total annual cost of $55.8 million for the associated minimum-cost solution. Figure 3.11 illustrates these effects.

We also must take into account MCO requirements for the AD/AFRC CRF network. Table 3.14 depicts the deployed CCRF manpower requirements necessary to support a range of deployment options, in a manner similar to the requirements presented in Table 3.10. However, whereas Table 3.10 presented the CRF requirements for support of the TF, Table 3.14 presents the CCRF and fixed network manpower

Figure 3.11
CRF Alternatives

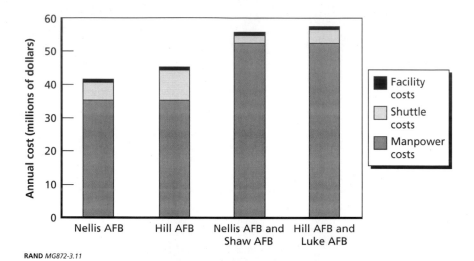

RAND *MG872-3.11*

Table 3.14
Alternative CRF Requirements to Support AD/AFRC

| Type of Network | Deployed CC Aircraft | | | | | |
	0% deployed	SSSP, 10% deployed	1 AEF, 20% deployed	2 AEF, 40% deployed	MCO, 80% deployed	100% deployed
Fixed	1,060	1,060	1,020	913	821	417
1 CCRF	—	208	285	460	879	1,069
2 CCRFs	—	330	416	570	920	1,072
3 CCRFs	—	369	495	753	987	1,149
4 CCRFs	—	492	660	832	1,140	1,212
5 CCRFs	—	615	825	1,040	1,340	1,425

required to support only AD and AFRC aircraft. Recall that, to support our illustrative steady-state scenario at a 4:1 dwell-to-deploy ratio, a total of 1,650 positions were needed, 330 at the CCRFs and 1,320 in the fixed network. Assume an illustrative MCO scenario similar to the one used for the TF analysis, requiring the capability to support 80 percent of the CC fleet deployed across two theaters, with no consideration given to deployment burden. Such a scenario for the AD/AFRC requires a total of 1,741 spaces, 821 in the fixed network and 920 to support two CCRFs. The difference between these 1,650 steady-state CRF positions and the 1,741 MCO CRF positions could be satisfied by using 91 AFRC drill positions. Note that this analysis does not address the adequacy of ANG maintenance manpower to support the deployment of its aircraft in this MCO scenario.

The assumption that deployed ANG forces are not supported by the CCRF network implies that deployed forces may operate under two different maintenance CONOPs. The AD and AFRC forces would have minimal maintenance on site and would rely on the repair network for most backshop and phase support, while deployed ANG units would have on-site ILM. It may be possible for the Air Force to have a single deployed maintenance concept without modifying ANG manpower levels if deployed ANG aircraft are to be supported via a CCRF, but the fixed CRF network does not support ANG home-station activities. Depending upon the level of ANG manpower contribution to the CCRF, this could potentially reduce the deployment requirement for AD/AFRC manpower, which would in turn allow a smaller AD/AFRC CRF manpower pool to support both home-station requirements and the steady-state scenario at an acceptable dwell-to-deploy ratio.

F-16 Overall Conclusions

Two alternatives have been identified for improving the effectiveness and efficiency of F-16 wing-level maintenance. In the first alternative, the Air Force can enhance operational effectiveness for the F-16 by adding a split-operations capability at each CC AS without increasing the baseline total maintenance manpower, whether the CRF network supports only the AD/AFRC forces (in which case, 850 maintenance positions are transferred into the AS to provide the split-operations capability) or the TF (in which case, the transfer of 1,900 maintenance positions into the AS is needed for split operations); note the almost identical totals at the bottom of Table 3.15.[26] In both cases, the manpower requirement associated with this split-operations "plus-up" can be captured by consolidating CRF workloads into a flexible maintenance network support concept. This CRF manpower is capable of supporting a long-term deployment of 10 percent of the CC fleet into two theaters and has a surge capability to support 80 percent of the CC fleet deployed into two theaters.

If, instead, the Air Force believes that its current F-16 maintenance operational capabilities are sufficient, an alternative policy would be simply to capture the savings associated with backshop centralization efficiencies and not add a split-operations capability to the squadrons.[27] As demonstrated by Figure 3.12, the Air Force would accrue

Table 3.15
Option 1: F-16 Increased Operational Effectiveness

Operation	Manpower Authorization		
	Current System	AD/AFRC-Only Repair Network	TF Repair Network
Group and MOS: FY 2008 UMD	3,363	3,363	3,363
AMXS: FY 2008 UMD	11,143	11,143	11,143
Moved from CMS and EMS		1,046	1,884
Split-operations plus-up		844	1,896
CMS and EMS			
Propulsion and avionics: FY 2008 UMD	2,863	2,863	2,863
Age and munitions: FY 2008 UMD	4,093	4,093	4,093
Phase and Related: FY 2008 UMD	6,221	2,714	
CRF network		1,741	2,532
Total	27,683	27,807	27,774

[26] In the middle data column, where only AD and AFRC manpower positions are rebalanced between the units and the repair network, the current 2,714 ANG positions assigned to phase and related backshops would not be modified.

[27] Alternatively, the Air Force might decide that, even though F-16 maintenance capabilities are stressed, these manpower savings would be better applied to some other career field. Another alternative for reducing manpower requirements would be to alter the deployment burden or RC participation policies discussed earlier.

Figure 3.12
Option 2: F-16 Increased Efficiencies

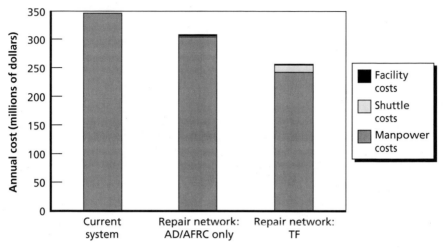

RAND *MG872-3.12*

savings whether the CRF network supported only the AD/AFRC forces or the TF, although the savings would be larger for the TF network. In either case, an economic rationale exists for repair network centralization. The bar on the left side of Figure 3.12 presents the manpower costs associated with the current system, including all CMS and EMS maintenance manpower for the shops under consideration. The center bar presents the total system costs for the CRF maintenance network alternative that supports only the AD and AFRC forces, with no split-operations capability added to the CC squadrons. The bar on the right side of the figure presents the total system costs for the TF CRF network alternative, again with no split-operations capability added to the CC squadrons. Under the current system, annual costs are $345 million, contrasted with $308 million for the AD/AFRC option ($37 million annual reduction) and $257 million for the TF option ($88 million annual reduction).

The manpower requirement dominates these costs. The manpower cost presented in Figure 3.12 includes AD, AFRC, and ANG for the CONUS, PACAF, and USAFE CRFs, supporting both the steady-state and MCO-surge requirements, along with those personnel who were previously in the CMS or EMS and are now reassigned to the AS, as well as the unchanged ANG phase and related backshop manpower for the AD/AFRC-only CRF network.[28] A relatively small shuttle cost is associated with aircraft movement between the aircraft operating locations and the CRFs.[29]

Recent large fluctuations in the price of aviation jet fuel led us to conduct additional analyses to identify how sensitive these alternative CRF network strategies were to variations in the shuttle cost. As a reference, the International Air Transport Association states on its Web site that, as of June 27, 2008, the price of aviation jet fuel increased 96 percent over the price one year earlier.[30] The F-16 CPFH used in this analysis was $6,500.[31] The CPFH includes many logistics costs in addition to aviation fuel, e.g., consumables, depot-level reparable assets, and depot maintenance costs. For the F-16C, aviation fuel constitutes $1,722, or 26 percent of the total CPFH. Because the shuttle costs are relatively small compared with the other costs presented in Figure 3.10, the AD/AFRC CRF network alternative would be less expensive than the current system even if the CPFH increased up to a factor of 13 times the $6,500 figure, or, holding all other CPFH components constant, if the price of aviation fuel increased up to a factor of 46 times the $1,722 figure. Similarly, the TF CRF network would be less expensive than the current system even if CPFH increased up to a factor of eight times the $6,500 value, or if the price of aviation fuel increased up to a factor of 28 times the $1,722 figure (holding all other CPFH components constant).

---

[28] We assumed an RC drill position personnel cost of 25 percent of the AD personnel cost of $65,000. For those RC positions that are assumed to be activated in support of steady-state deployed operations, we assumed an additional personnel cost of $65,000, equal to the AD personnel cost.

[29] This shuttle cost is presented only for home-station operations because of the uncertainty associated with deployed operating locations.

[30] Jet fuel prices from International Air Transport Association, n.d.

[31] This figure was based on U.S. Air Force, 2006, Table A4-1; the precise CPFH values given in this reference vary by F-16 series, with the F-16C, the most common series in the inventory, having a CPFH of $6,543.05 (the F-16A and F-16B had slightly lower CPFHs, and the F-16D had a slightly higher CPFH).

The facility costs associated with the establishment of CRFs are also presented for the maintenance network alternatives; however, they amount to a small fraction of the total annualized costs. This suggests that, even though the facility costs presented in this analysis are likely to be somewhat underestimated, even if they were understated by a factor of 10, they would not be so large as to have a material effect on the conclusions.

The Air Force could also choose to implement an alternative lying between the two endpoints of "enhanced effectiveness" and "increased efficiency" for F-16 maintenance. For example, it could select a posture that adds a split-operations capability to some, but not all, CC squadrons, if it wished to capture some effectiveness increases while also allowing some reallocation of resources to career fields other than aircraft maintenance.

CHAPTER FOUR

Alternatives for Rebalancing KC-135 Maintenance Resources

The analysis detailed in Chapter Three suggested that a realigning of F-16 backshop maintenance capability, from the current decentralized "unit self-sufficient" posture to a centralized "maintenance network," could allow reductions in backshop maintenance manpower with no decrease in backshop support to the flying unit. Alternatives would then exist to either (1) create a new split-operations capability via a transfer of these positions into F-16 sortie-generation maintenance; (2) free up manpower resources for other, more stressed, career fields outside of aircraft maintenance; (3) capture the savings associated with this reduction in backshop manpower; or (4) some combination of the above. To determine whether the maintenance network concept affords similar benefits to other weapon systems, we performed a similar analysis on a dissimilar MDS: the KC-135. Like the F-16 analysis, this analysis focuses on rebalancing KC-135 maintenance resources between the flightline and the backshops. We will again consider both a TR alternative and an alternative that rebalances only AD and AFRC resources with the repair network.

Determination of KC-135 AS Maintenance Tasks and Manpower Requirements

Table 4.1 shows the end-of-FY 2008 manpower authorizations associated with KC-135 wing-level maintenance, as presented in UMDs across the

Table 4.1
KC-135 Maintenance Personnel

| Operation | AD | Manpower Authorization | | | | |
| | | ANG | | AFRC | | |
		Part-Time	Full-Time	Part-Time	Full-Time	Total
Group and MOS	483	400	309	159	191	1,542
AMXS	2,167	758	585	676	436	4,622
MXS	1,427	1,880	1,471	365	430	5,573
Total	4,077	3,038	2,365	1,200	1,057	11,737

SOURCE: KC-135 FY 2008 UMD.

NOTE: This table presents manpower authorizations, which typically exceed the actual manpower assigned to any organization. Because this analysis focuses on maintenance manpower requirements, we present all manpower in terms of authorization levels.

AD, ANG, and AFRC. This set of manpower authorizations comprises approximately 12,000 positions, located in the following organizations: maintenance group, MOS, AMXS, and MXS.[1] For the KC-135, the scope of this analysis includes the maintenance manpower in the AMXS, which performs most of the flightline maintenance workload, and the MXS, which performs most of the backshop maintenance workload. Unlike the F-16 analysis, from which we excluded a number of backshops, our KC-135 analysis includes all maintenance manpower positions in these two squadrons, approximately 10,000 total UMD positions.

For the F-16, it was necessary to perform LCOM analyses to identify the maintenance workloads that must remain at the aircraft operating location and those that could be performed at a centralized off-site facility. Such analysis had already been completed for the KC-135 in support of the Air Mobility Command (AMC) FOL/regional maintenance facility (RMF) construct for deployed AMC operations (see U.S. Air Force, 1999). In the present analysis, we took that AMC deployment construct and applied it to home-station operations, iden-

---

[1] See Appendix A for more details of this manpower analysis.

tifying the work centers as presented in Figure 4.1.[2] The AS work centers, shown on the left side of the figure, include the AMXS shops appearing at the top of the list, along with the split shops (indicated by an "–AS" after the shop name) for which there is a capability at both the AS and the CRF.

Because AMC has incorporated these maintenance constructs into its deployment concepts, it was not necessary to perform new LCOM runs to identify the AS manpower requirements. Instead, we took the Unit Type Code (UTC) deployment requirements as identified for KC-135 maintenance and applied them to all home-station

Figure 4.1
KC-135 AS and CRF Work Centers

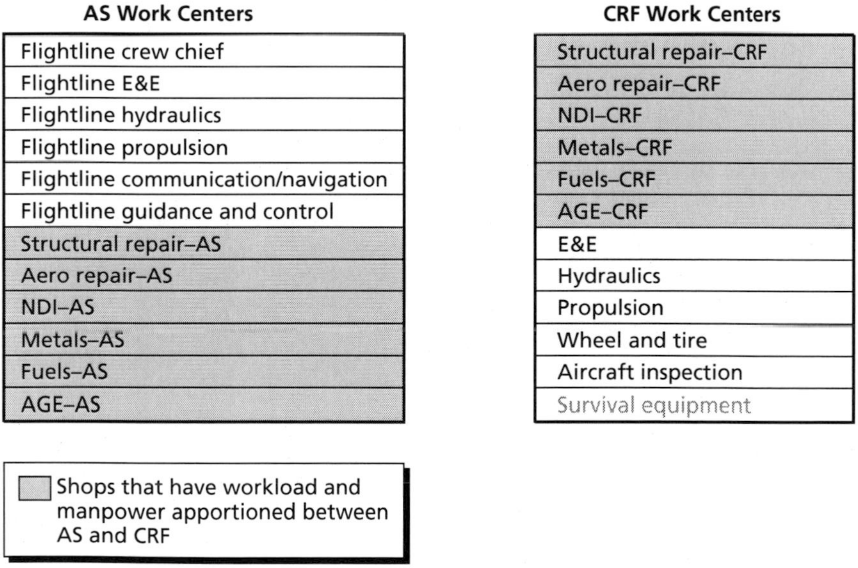

AS Work Centers
Flightline crew chief
Flightline E&E
Flightline hydraulics
Flightline propulsion
Flightline communication/navigation
Flightline guidance and control
Structural repair–AS
Aero repair–AS
NDI–AS
Metals–AS
Fuels–AS
AGE–AS

CRF Work Centers
Structural repair–CRF
Aero repair–CRF
NDI–CRF
Metals–CRF
Fuels–CRF
AGE–CRF
E&E
Hydraulics
Propulsion
Wheel and tire
Aircraft inspection
Survival equipment

Shops that have workload and manpower apportioned between AS and CRF

RAND MG872-4.1

---

[2] As with the F-16 analysis, since the publication of the AMC LCOM report, the survival equipment work center was moved from the maintenance organization to the operations organization. Thus its manpower has been excluded from this analysis.

units across the TF (AD, ANG, and AFRC).[3] These UTCs contain maintenance manpower requirements for the AS work centers listed in Figure 4.1. The manpower requirements obtained in this analysis are presented in Table 4.2. The top row shows the FY 2008 manpower authorization totals for KC-135 AMXS. The next two rows present the AS manpower alternative, using a UTC-based AMXS requirement, differentiating between AS manpower formerly assigned to the AMXS and that formerly assigned to the MXS (which would be operating as split shops). Approximately 1,400 maintenance manpower positions that were previously in KC-135 backshops would be moved to the AS under this concept. As with the F-16, we further specified a split-operations maintenance manpower requirement for the KC-135 AS, identifying by UTC the additional manpower requirements to support each CA squadron operating one half of its PAA on deployment and the other half at home station. This KC-135 split-operations capability requires approxi-

Table 4.2
Manpower Requirements for KC-135 AS Operations

| | | Manpower Authorization | | | | |
| | | ANG | | AFRC | | |
Operation	AD	Part-Time	Full-Time	Part-Time	Full-Time	Total
AMXS FY 2008 UMD	2,167	758	585	676	436	4,622
UTC-based: AMXS	1,960	1,152	889	506	326	4,833
UTC-based: moved from MXS	497	353	272	148	96	1,366
Split-operations plus-up	789	651	502	258	166	2,366
Proposed new AS	3,246	2,156	1,663	912	588	8,565

---

[3] The following UTCs were used for this analysis: for the KC-135E, HFKLE, HFK4E, HFK2B, and HFK1E; for the KC-135R, HFKLR, HFK4R, HFK2A, and HFK1R. For all CA KC-135 squadrons, we used the UTC requirement associated with the squadron PAA as its manpower requirement. To determine home-station manpower requirements (for the training squadron at Altus AFB and for the development of split-operations requirements), we developed planning factors to vary the UTC requirement according to the reduced OPTEMPO at home station and the decreased man-hour availability factor (MAF) at home station. See Appendix E for more details on these computations.

mately 2,400 additional maintenance manpower positions beyond what is required for each squadron to operate in a fully deployed scenario.

KC-135 Repair Network Design Options

We now turn to evaluation of the KC-135 network of CRFs.

To identify how networked CRF maintenance affects the logistics system performance, we again need to determine aircraft inspection production times and component pipeline times and requirements. Whereas F-16s have a scheduled phase inspection that is performed at intervals defined by the aircraft's accumulated flying hours, KC-135s have a periodic inspection (PE) that is performed at an interval determined by the first of two criteria to be satisfied: flying hours since last PE (currently 1,500) or calendar months since last PE (currently 15).[4]

To identify the amount of time an aircraft would be unavailable due to PE, we need to determine the PE throughput time, excluding all hours during which no maintenance occurs. Figure 4.2 presents such throughput times for the KC-135 fleet, using REMIS data over the interval 2004–2007, differentiated by MAJCOM or component. We used a process similar to the one employed for the F-16 analysis to obtain the data presented here; for more detail on the analytic procedure, see Appendix C. The x-axis in Figure 4.2 presents the number of PE throughput days. The y-axis presents the number of occurrences for each PE throughput, by MAJCOM or component, over this three-year interval. Having removed those hours during which no maintenance occurred, the shape of this throughput time distribution is roughly similar across MAJCOMs and components, with mean PE throughput times ranging from eight to 14 days (with exception of the ANG, whose mean throughput time is 16 days).

In the remainder of this analysis, we assume a 14-day KC-135 PE throughput interval. We also assume 24-hours-per-day, seven-days-per-

---

[4] These intervals have been extended in recent years. The KC-135 fleet is currently undergoing testing to determine whether they can be extended to 1,800 flying hours and 18 months.

Figure 4.2
Throughput Times for KC-135 Fleet

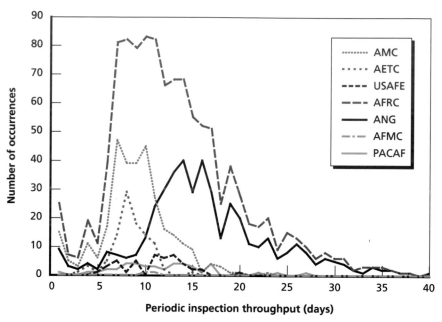

RAND *MG872-4.2*

week CRF operations, with a three-shifts-per-day, 40-hours-per-week-per-person factor during steady-state operations at home station.[5]

As with the F-16 analysis, we attempted to identify the effect of networked KC-135 CRF maintenance for home-station operations on the movement of failed aircraft components between the aircraft operating location and the CRF. Focusing, as before, on the set of aircraft components appearing in both the current RSP for any Air Force KC-135 unit and the RAND March 2006 capture from the D200 RDB data system across all KC-135 series, we identified a set of 298 unique NIINs. For more details on this pipeline analysis, see Appendix F.

---

[5] Were a less-stressing work schedule assumed at home station, the home-station aircraft availability would be expected to decrease in a manner consistent with the values presented for the F-16 in Table 3.7, although the manpower requirements would not be expected to change significantly across different KC-135 work schedules.

For this KC-135 transportation and inventory analysis, we further restrict our focus to that fraction of component failures that were previously repaired at the on-site backshops.[6] Assuming a notional home-station daily flying schedule of 1.75 flying hours per PAA (for both CONUS-based aircraft and permanently assigned PACAF and USAFE forces) for the set of NIINs under consideration and counting only those failures that would currently be repaired on-site, we would expect to observe a daily fleetwide average of 73.0 component failures. The expected annual fleetwide transportation cost associated with this notional home-station scenario is $2.4 million, assuming, as we did in the F-16 analysis, that all failed components are shipped using the cost structure associated with FedEx Small Package Express two-day rates for U.S. domestic shipments.

An additive inventory requirement would also be needed to support the new transportation segments introduced by the CRF. The two-day transport time, in each direction, that is assumed above results in a requirement for four days' worth of pipeline inventory for CONUS aircraft. We assume that permanently assigned aircraft outside the continental United States (OCONUS) would be supported from this same inventory pool, with a 14-day transport time to CONUS, in each direction,[7] generating a 28-day pipeline requirement for OCONUS units. If we assume that a single inventory requirement is computed to support the worldwide KC-135 fleet, operating at the notional flying schedule of 1.75 flying hours per day, a total one-time inventory investment of $6.8 million would be required to support home-station operations.

---

[6] The assumption that the transportation of items between the operating unit and the depot would not be affected under this option would be valid as long as the remaining organizational-level maintenance could identify those items that require depot maintenance.

[7] We performed an analysis of Military Aircraft Issue Priority Group 1 (IPG1) shipments for the first ten months of 2007, using the RAND-maintained Strategic Distribution Database. This database aggregates defense-related pallet movement data feeds, including the AMC Global Air Transportation Execution System (GATES) database. This analysis suggested an average travel time of 14 days from the CONUS to each of U.S. European Command (EUCOM) and Pacific Command (PACOM).

Because this inventory requirement is a one-time additional investment, this cost could be amortized across the expected duration of KC-135 CRF operations. Considering an amortization interval as short as seven years produces an annualized inventory requirement cost of less than $1 million. Further, note that a transportation pipeline and inventory requirement would not necessarily be created for every unit, e.g., if the CRF were located at an existing KC-135 operating location. Thus, because the costs associated with CRF component repair transportation and inventory are relatively small, we do not include them in the remainder of this analysis, focusing instead on the other, much larger system costs.

Unlike the F-16, for which RAND had to conduct its own LCOM analyses to identify CRF manpower requirements, we were again able to draw on an existing AMC analysis. U.S. Air Force (1999) identified maintenance manpower requirements for RMFs supporting a range of PAA, under a sustained wartime tasking. We built upon this analysis, developing regression-based planning factors that we used to translate these deployment requirements into the manpower positions necessary to support home-station operations. For more details on this calculation procedure, see Appendix E.

Based on our extension of these AMC KC-135 LCOM analyses, it appears that the KC-135 also demonstrated strong labor economies of scale for CRF maintenance, as shown in Figure 4.3. The x-axis presents the number of PAA operating in a home-station environment supported at a CRF. The y-axis presents the CRF maintenance manpower authorizations required per PAA. The general behavior is the same as that observed in the F-16 analysis: At the left end of the curve, a large relative requirement for a small number of PAA supported; at the right end of the curve, a small relative requirement for a large number of PAA supported; and a point at which the curve flattens at approximately 300 PAA, beyond which we observe no additional marginal reduction in maintenance manpower for increased numbers of supported PAA. Interestingly, there is again a proportional decrease of roughly 3:1 between the manpower requirements for the smallest and largest units, as was observed for the F-16.

Figure 4.3
KC-135 CRF Manning Requirements, Home Station

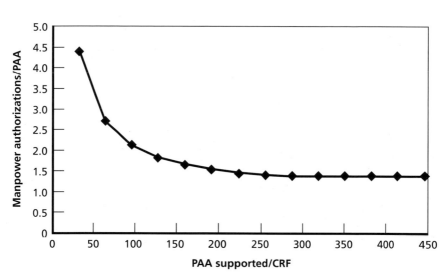

RAND *MG872-4.3*

CRF Network to Support Contingency Operations

We now turn our attention to the design of the CRF network. We again begin the analysis by focusing on the requirements of the contingency-deployed fleet. We considered only one option for providing CRF support to the KC-135 deployed fleets: Aircraft and components are retrograded to a fixed network site for CRF maintenance. We eliminated the FOL-unique support option and contingency CRF options from the analysis, since the KC-135's flying range and long maintenance intervals make a reachback option wherein KC-135s receive CRF maintenance from a fixed network site a potential support option for any deployment location. For this reason, we considered the requirements of the KC-135 flexible and fixed networks simultaneously.

Because we do not consider a forward-deployed CRF maintenance option for the KC-135, the risks associated with a potential attack on backshop manpower at each FOL or at a centralized in-theater CRF are minimized, and the deployment requirements associated with KC-135 units are decreased accordingly. However, the reliance on off-site support via connectivity to CRFs in the fixed repair network imposes risks associated with potential disruptions to the transportation network that links FOLs to CRFs. The very long KC-135 maintenance intervals, which would likely exceed the duration over which any individual aircraft or aircrew would be deployed, make this risk small for KC-135 aircraft inspection activities.

The optimization approaches we developed for the F-16 analysis were also used for the KC-135.[8] The parameters driving the design of the KC-135 network are similar to those used for the F-16:

- size of the deployed fleet (PAA)
- intensity of OPTEMPO (UTE rate and ASD)
- duration of deployment (for contingency forces)
- geographic dispersion (for both home-station and contingency operating locations)
- extent of labor scale economies
- minimum manpower needed to staff CCRFs
- personnel costs ($65,000 per man-year)
- aircraft shuttle cost ($5,370 per flying hour; costs associated with shuttle to and from deployed FOLs are not included)
- facility costs ($1 million per year per PE dock amortized).

The aggregate flying hours to be supported define the requirement for component repair and periodic inspection workloads (although KC-135 PE also depends on calendar time since last inspection). For home-station aircraft, the FY 2008 programmed flying hours are again used to determine the flying-hour requirement. Personnel and aircraft shuttle costs were determined as before,[9] using a personnel cost of $65,000 per manpower authorization, a KC-135 CPFH of $5,370, and a KC-135 block-speed planning factor of 420 nmi/hr (U.S. Air

---

[8] See Appendix D for more details on these optimization models.

[9] Recall that establishing CRFs might also incur costs for low-inventory equipment in split shops (see the F-16 discussion for further detail).

Force, 2006). As in the F-16 analysis, the construction of new aircraft inspection docks at the CRFs generated the only facility cost under consideration; note that the annualized facility costs are significantly higher for the KC-135 than those for the F-16 ($1 million compared with $100,000), primarily because of the larger size of the aircraft and the associated larger hangar-space requirements.[10]

Again, as we did for the F-16, we focus on the performance of CRF alternatives with respect to their total system costs, setting aside operational metrics, because we assume that (1) all CRF alternatives identified will have sufficient maintenance manpower to accomplish all required workloads without lowering operational performance due to the buildup of large maintenance queues, and (2) the operational effect of flying home-station aircraft between their operating location and a PE CRF is unclear, since the aircraft are, by definition, "mission capable" during this sortie, which could potentially be used for training missions. For deployed aircraft, the relatively long PE interval (15 months) suggests that, provided only "clean" aircraft are deployed, no KC-135 should require a dedicated sortie from the deployed location to a CRF, since aircraft would typically return to home station within a 15-month interval for reasons other than maintenance. However, we again assume that the shuttle cost between home station and the PE CRF is a 100 percent additional cost, which is equivalent to saying that no training missions are accomplished during this sortie. As was the case with the F-16, this assumption makes centralization less attractive and would lead to less-centralized networks being preferred, to the extent that the shuttling costs are relatively large.

As before, we analyzed these maintenance manpower requirements to address the capability of the maintenance force to support a variety of potential deployment requirements. Rather than suggesting that the Air Force build its maintenance manpower capability to any specific level, we present an exemplary force-sizing analysis built

---

[10] This cost was obtained in a manner similar to that described for the F-16 in Chapter Three. Recall (1) the assumption that new hangars to support PE maintenance would be required at each potential CRF, and (2) that ACC/A4 staff suggested that other important facility costs (primarily for a fuel barn) might be missing from this analysis.

around a notional desired capability level. The analytic process that follows could be applied to any other desired capability level. We note again, however, that the OSD guidance discussed in Chapter One directs the services to plan and program for a future in which some fraction of the force is continuously deployed forward.

For the purposes of illustration, assume that the Air Force desires the capability to maintain a steady-state deployment of 40 percent of the CA aircraft into two theaters, with 20 percent in each theater.[11] This steady-state deployment requirement is significantly higher than the 10 percent we assumed for the F-16. This level is broadly consistent with OSD guidance, but once again, this is presented as an illustration. A number of the policy options that were considered in the F-16 analysis do not apply to the KC-135 analysis (e.g., acceptable dwell-to-deploy ratios, fraction of the forward-deployed manpower assigned to the AD), since we have assumed that all KC-135 CRF support is provided by the fixed-network sites; recall the previous assumption that in the steady state, all fixed-network CRF personnel work on a 40-hour-per-week schedule.

Because of the long flying range of the KC-135, we did not consider the USAFE and PACAF units separately from the CONUS units, but rather considered support to the worldwide KC-135 fleet (including the 40 percent deployment across two theaters utilizing reachback to the fixed network, as mentioned above). This worldwide KC-135 beddown, including AD, AFRC, and ANG units, is shown in Figure 4.4.

The optimization model identifies the minimum-cost network as having two CRFs: one in the central United States (at Scott AFB) and one in the western United States (at Fairchild AFB). Table 4.3 presents the performance of a set of CRF alternatives, similar in format to the F-16 information presented in Table 3.9. Cost and manpower details are presented for the minimum-cost networks, as identified by the

---

[11] We have not assumed that the deployment requirement varies across future years, as presented in Figure 2.1. Rather, we have identified a capability level that can sustain a given level of deployment activity across the entire steady-state planning horizon.

Figure 4.4
FY 2008 KC-135 Worldwide Beddown

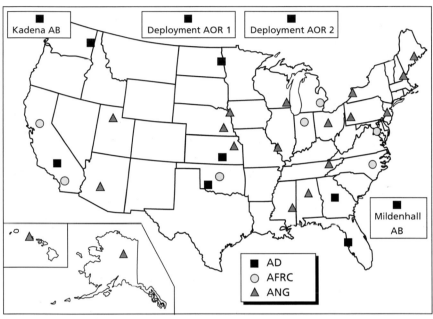

RAND *MG872-4.4*

Table 4.3
KC-135 CRF Network Options

Item	Costs and Manpower Positions				
	1 CRF	**2 CRFs**	**3 CRFs**	**4 CRFs**	**5 CRFs**
Manpower costs ($M/year)	66.4	65.9	67.4	74.6	83.9
Shuttle costs ($M/year)	5.4	4.3	3.5	2.6	2.4
Facility costs ($M/year)	19.0	19.0	19.0	19.0	19.0
Total annual costs ($M/year)	90.7	89.2	89.9	96.2	105.4
Manpower positions	1,021	1,014	1,037	1,148	1,291

optimization model, containing between one and five CRFs. Shuttle costs from the deployed locations to the fixed network are not included here because of uncertainties associated with the precise locations of these deployments. However, these networks have sufficient manpower and facilities to support both the home-station and deployed aircraft in this steady-state scenario. Note that the total annual costs are relatively constant across the one-, two-, and three-CRF network solutions.

Figure 4.5 contrasts the performance of the best single-CRF, two-CRF, and three-CRF networks. The three options have similar total costs, indicating that the KC-135 network is relatively insensitive to the number of CRF locations established across this range. The manpower requirement is essentially constant across these three solutions; the small variations are due to the "chunky" manner in which manpower is added on an integer per-aircraft-inspection-space basis. The facility requirements are also constant across the three solutions, because they are applied on a per-aircraft-inspection-space basis, and each solution

Figure 4.5
Comparative Cost of KC-135 Options

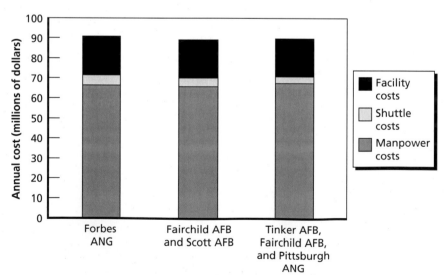

constructs the minimum number of spaces necessary (19) to support the KC-135 PE workloads assumed in this scenario.[12] The three-CRF network generates the smallest shuttle cost across these three solutions because the average operating-location-to-CRF distance is less for a three-CRF network than that for either a single- or two-CRF network. Figure 4.5 provides a graphical comparison of the costs associated with these alternatives, again showing KC-135 CRF support to be insensitive to the precise number of locations established. It is possible to establish one, two, or three CRF locations with little effect on cost performance.

Figure 4.6 adds a fourth network alternative to the set presented in Figure 4.5, contrasting the performance of an alternative three-CRF

Figure 4.6
Alternative KC-135 Options

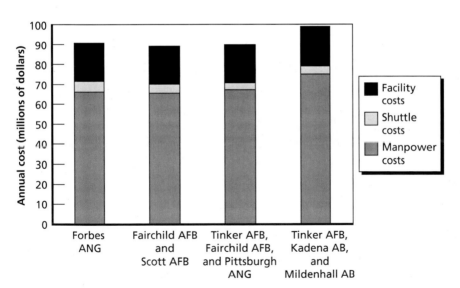

RAND *MG872-4.6*

---

[12] For the home-station and deployed aircraft that are supported in this scenario, there is a total annual requirement for 458 periodic inspections; assuming a throughput of 14 days per inspection and 350 working days per year, there is a total requirement for 19 KC-135 PE hangar spaces.

solution with a single CONUS CRF at Tinker AFB and two OCONUS CRFs at Kadena and Mildenhall Air Bases (ABs).[13] The primary difference in cost between this solution and those presented in Figure 4.5 is that of manpower: The CRFs at Kadena AB and Mildenhall AB are supporting only their collocated units and some fraction of the deployed aircraft. These OCONUS sites are small and therefore achieve substantially less manpower scale economy. However, the total cost for this solution is comparable to that of the overall minimum-cost solution, suggesting that, as with the F-16, there is further relative insensitivity to the precise CRF locations selected for the KC-135 CRF maintenance network. As discussed previously, this allows a range of other considerations beyond the scope of this analysis to enter into the final CRF location decision. For example, a CRF at Tinker AFB might provide proximity to the depot, while permanent USAFE and PACAF CRFs might improve support to certain deployed forces.

The total steady-state manning requirement can be determined for any desired network. Consider the three-CRF solution with OCONUS CRFs at Kadena AB and Mildenhall AB, as presented above. The total steady-state manpower for this solution consists of 1,160 maintenance manpower positions.

In addition to performing an analysis of steady-state requirements, it is also necessary to consider KC-135 requirements for other potential scenarios, such as MCOs. By way of illustration, suppose a notional MCO scenario involves the deployment of 100 percent of the CA KC-135 fleet into two theaters, with 50 percent deployed to each. Assume that all deployed forces will still reach back to the fixed network for CRF support but that CRF manpower now works under wartime man-hour availability factors, with 60-hour workweeks per person. The deployment requirement is then 1,065 total positions. This is less than the three-CRF KC-135 steady-state manpower requirement of 1,160 positions determined earlier, suggesting that no additional drill positions would be needed to support surge requirements. However,

---

[13] Manpower costs were assumed to be identical for permanently stationed CRF positions at CONUS and OCONUS locations, with no adjustment made for overseas cost-of-living allowance or overseas housing allowance.

while the increased utilization of CRF maintenance manpower deriving from the change from a 40-hour workweek to a 60-hour workweek was sufficient to allow all MCO workloads to be accomplished with no increase in manpower over the steady-state manpower requirement, we have assumed that all CRF facilities are being operated on a 24/7 basis in the steady state. Thus, additional CRF facilities need to be procured to accommodate the MCO surge in demand, at an additional total annualized cost of $6 million.

Of course, all these requirements are a function of the scenarios selected and the policy choices implemented. Because we have selected a fairly high-stress steady-state scenario for the KC-135, the steady-state manpower requirement we calculate exceeds MCO requirements. As discussed earlier, the force-sizing analysis presented in this section is not intended to represent a recommended capability level (although it is broadly consistent with OSD planning and programming guidance). Rather, it is meant to illustrate how the manpower requirements that were identified in this analysis can be translated into a total maintenance force-sizing construct, with the selection of a few additional policy choices, such as number of CRFs desired. This analytic process could be applied to any other capability level the Air Force deemed appropriate.

CRF Networks to Support Only AD and AFRC Forces

As was done for the F-16, we performed a KC-135 analysis that assumed the repair network construct supported only AD and AFRC units, with ANG units retaining their current maintenance construct. For this analysis, we assumed that AD and AFRC units receive both home-station and deployed support from the fixed CRF network. We assumed that ANG aircraft are not supported by the repair network, either while at home station or while deployed, but all deployed ANG aircraft are retrograded to receive CRF component repair and aircraft inspection at the home-station unit.[14] Whereas the previous TF analyses reallocated manpower among all units and the repair network, these additional analyses reallocated only AD/AFRC manpower within the network; ANG manpower authorizations were not modified. We followed the analytic process utilized earlier.

Baseline Maintenance Manpower

First, we determined the baseline maintenance manpower. Table 4.4 shows the FY 2008 UMD authorizations for AD, ANG, and AFRC KC-135 maintenance manpower; the ANG manpower is grayed out, since it is not modified in this additional analysis. The AMX and MXS numbers are again highlighted because the reallocation of positions between these categories remains the focus of the analysis. Note that 54 percent of the KC-135 maintenance manpower is in the AD and AFRC components.

AS Requirements

As with the F-16, because the KC-135 AS manpower requirement was computed on the basis of individual squadrons, it is not difficult to identify the manpower necessary to create an AS at each squadron and to add a split-operations capability at each CA squadron for the AD/AFRC only. Table 4.5 modifies the results presented in Table 4.2, indicating that the establishment of an AS maintenance capability at AD and AFRC KC-135 squadrons requires a total of approximately 4,700 positions. The UTC-based AMXS requires approximately 500 fewer positions than currently exist in AD/AFRC AMXS UMDs. There is a reassignment of roughly 750 positions that were previously in the MXS backshops into the AS and a new requirement for approximately 1,200 split-operations positions that do not currently exist in maintenance UMDs. Because we have assumed that ANG manpower is not modified in this analysis, all ANG manpower columns are grayed out, and ANG flightline operations are assumed not to receive the additional

---

[14] Because of the assumption that all deployed KC-135 receive CRF support via reachback, the complications associated with multiple deployed maintenance concepts are not relevant here.

Table 4.4
AD and AFRC KC-135 Maintenance Manpower Authorizations

| | | Manpower Authorization | | | | | |
| | | ANG | | AFRC | | | |
Operation	AD	Part-Time	Full-Time	Part-Time	Full-Time	Total	Total AD/AFRC
Group and MOS	483	400	309	159	191	1,542	833
AMXS	2,167	758	585	676	436	4,622	3,279
MXS	1,427	1,880	1,471	365	430	5,573	2,222
Total	4,077	3,038	2,365	1,200	1,057	11,737	6,334

SOURCE: KC-135 FY 2008 UMD.

Table 4.5
Manpower Requirements for AD and AFRC KC-135 AS Operations

| | | Manpower Requirement | | | | | |
| | | ANG | | AFRC | | | |
Operation	AD	Part-Time	Full-Time	Part-Time	Full-Time	Total	Total AD/AFRC
AMXS FY 2008 UMD	2,167	758	585	676	436	4,622	3,279
UTC-based: AMXS	1,960	1,152	889	506	326	4,833	2,792
UTC-based: moved from MXS	497	353	272	148	96	1,366	741
Split-operations plus-up	789	651	502	258	166	2,366	1,213
Proposed new AS	3,246	2,156	1,663	912	588	8,565	4,746

split-operations manpower computed earlier. The relative manpower increase associated with split operations was larger for the ANG (86 percent of AMXS) than for either the AD or the AFRC (36 and 38 percent, respectively). As with the F-16, this implies that an AD/AFRC-only CRF network will need a smaller reduction in backshop positions as a result of centralization to realize a constant maintenance manpower total with the current baseline.[15]

---

[15] This set of assumptions again leads to the creation of squadrons of differentiated capabilities, with AD and AFRC squadrons staffed to support split operations and ANG squadrons lacking this increased capability.

We used the optimization model described previously to identify alternatives for the fixed CRF network, now supporting only AD
and AFRC forces. As indicated by Figure 4.7, the worldwide KC-135
network becomes much simpler with the exclusion of all ANG units,
reducing the total number of locations to one-half the number in the
TF network. Recall that we did not consider the USAFE and PACAF
units separately from the CONUS but rather considered support to the
worldwide KC-135 fleet (including the steady-state scenario's deployment into two theaters that are supported via reachback to the fixed
network).

The optimization model identified a minimum-cost solution that
establishes one CONUS CRF at McConnell AFB. However, as was
observed in the TF analysis, the AD/AFRC-only CRF network exhibits relative insensitivity to both the *number of CRFs* and their *precise
locations*, allowing considerations outside the scope of this analysis to

Figure 4.7
FY 2008 AD and AFRC KC-135 Worldwide Beddown

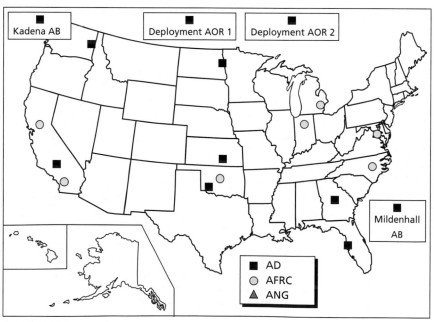

factor into the CRF network design decisions without incurring large cost effects. Suppose that, as before, a desire for proximity to the depot argued for the establishment of a KC-135 CRF at Tinker AFB. As another alternative, suppose that the potential to better support deployed forces argued for the establishment of fixed CRF sites in PACAF and USAFE. Table 4.6 presents further details on the costs associated with the minimum-cost solution and these two alternatives.

The one-CRF alternatives differ only by a very slight variation in shuttle costs; moving the CRF from McConnell AFB to Tinker AFB has almost no discernible impact on total costs. If establishing a permanent CRF capability in PACAF and USAFE were desired, a third CRF would be required in the CONUS (selected here as Tinker AFB), according to the optimization model. Contrasting this three-CRF network with the minimum-cost solution, the three-CRF network's ability to potentially improve support to deployed forces via the establishment of CRFs in PACAF and USAFE generates an associated annual personnel cost increase of $10.1 million, an annual reduction of $1.4 million in shuttle costs, and an annualized increase of $1 million in facility costs, for a net annual increase of $9.7 million and an associated increase of 156 manpower positions. This manpower increase is due to the substantial reductions in scale economy resulting from distributing the total amount of work across three sites. As before,

Table 4.6
Costs for KC-135 AD/AFRC Fixed-Network Options

| Item | Costs and Total Manpower Positions | | |
	1 CRF, McConnell AFB	1 CRF, Tinker AFB	3 CRFs, Tinker AFB, Kadena AB, Mildenhall AB
Manpower costs ($M/year)	46.8	46.8	56.9
Shuttle costs ($M/year)	2.8	2.9	1.4
Facility costs ($M/year)	13.0	13.0	14.0
Total annual costs ($M/year)	62.6	62.7	72.3
Total manpower positions	720	720	876

shuttle costs from the deployed locations to the fixed network are not included because the precise locations of these deployments are uncertain. Figure 4.8 illustrates these points.

As with the TF KC-135 CRF analysis, the AD/AFRC-only CRF analysis determined that, for an MCO scenario in which 100 percent of CA KC-135s are deployed into two theaters, with all AD/AFRC CRF workload performed via the fixed network, the MCO requirement of 610 positions is less than the steady-state CRF manpower requirement. Again, this is primarily because (1) the steady-state scenario is rather taxing (40 percent of the CA aircraft are deployed) and (2) the change from a 40-hour workweek during the steady state to a 60-hour workweek during an MCO generates significantly more capability from the steady-state manpower pool. This analysis does not address the adequacy of ANG maintenance manpower to support the deployment of its aircraft in this MCO scenario. Note also that, because CRF facilities are assumed to operate 24/7 during the steady state, an additional $1 million in annualized facility costs is

Figure 4.8
CRF Options

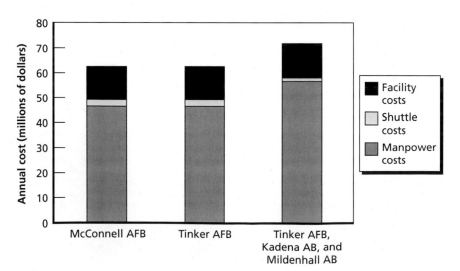

necessary for the AD/AFRC-only CRF network to build the necessary facility surge capacity to accommodate MCO workloads.

KC-135 Overall Conclusions

This analysis identified two alternatives for improving the effectiveness and efficiency of KC-135 wing-level maintenance, similar to the results presented for the F-16. In both of these alternatives, the existing AMC FOL/RMF construct that is used for maintenance support to deployed forces is applied to home-station operations as well. In the first alternative, the Air Force can enhance operational effectiveness for the KC-135 by rebalancing current maintenance resources with the addition of a split-operations capability at each CA AS without increasing the baseline total maintenance manpower, whether the CRF network supports only the AD/AFRC forces (in which case, 1,200 maintenance positions are transferred into the AS to provide the split-operations capability) or the TF (in which case, the transfer of 2,400 maintenance positions into the AS is needed for split operations); the totals, shown at the bottom of Table 4.7, are almost identical. In both cases, the man-

Table 4.7
Option 1: KC-135 Increased Operational Effectiveness

Operation	Manpower Authorization		
	Current System	AD/AFRC-Only Repair Network	TF Repair Network
Group and MOS, FY 2008 UMD	1,542	1,542	1,542
AMXS: FY 2008 UMD	4,622	1,343	
UTC-based AMXS		2,792	4,833
UTC-based moved from MXS		741	1,366
Split-operations plus-up		1,213	2,366
MXS			
FY 2008 UMD	5,573	3,351	
CRF network		876	1,160
Total	11,737	11,858	11,267

NOTE: In the middle data column, where only AD and AFRC manpower positions are rebalanced between the units and the repair network, the current ANG manpower authorizations of 1,343 AMXS positions and 3,351 MXS positions would not be modified.

power associated with this split-operations "plus-up" can be captured by consolidating CRF workloads into a flexible maintenance network support concept. This CRF manpower is capable of supporting a long-term deployment of 40 percent of the CA fleet into two theaters and has a surge capability to support 100 percent of the CA fleet deployed into two theaters.

Alternatively, if the Air Force believes that its current KC-135 maintenance operational capabilities are sufficient, it could decide simply to capture the savings associated with backshop centralization efficiencies and not add a split-operations capability to the squadrons.[16] As shown in Figure 4.9, the Air Force would accrue savings whether the CRF network supported only the AD/AFRC forces or the TF, although the savings would be larger for the TF network. There is an economic rationale for repair network centralization in both cases. The bar on the

Figure 4.9
Option 2: KC-135 Increased Efficiencies

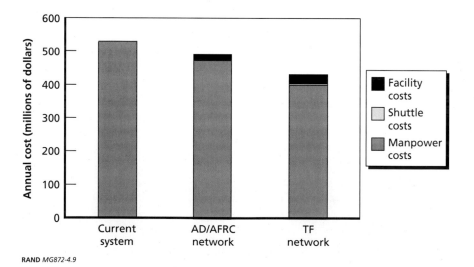

RAND *MG872-4.9*

---

[16] Alternatively, the Air Force might decide that, even though KC-135 maintenance capabilities are stressed, these manpower savings would be better applied to some other career field.

left side of Figure 4.9 presents the manpower costs associated with the current system, including all AMXS and MXS manpower. The analysis did not assume that the KC-135 AMXS UMD was held constant (as was assumed for the F-16); instead, all AS manpower (including the AMXS component) was redesigned around UTC requirements.[17] Thus the AMXS manpower must be included in any presentation of potential manpower reductions. The center bar presents the total system costs for the CRF maintenance network alternative that supports only the AD and AFRC forces, with no split-operations capability added to the CA squadrons. The bar on the right side of the figure presents the total system costs for the TF CRF network alternative, again with no split-operations capability added to the CA squadrons. Under the current system, annual costs are $531 million, contrasted with $488 million for the AD/AFRC option ($43 million annual reduction) and $429 million for the TF option ($102 million annual reduction).

As with the F-16, the total costs are dominated by the manpower requirement. The manpower cost presented here includes AD, ANG, and AFRC for all AMXS, MXS, AS, and CRF positions capable of supporting both the steady-state and MCO scenarios.[18] There is a small shuttle cost associated with aircraft movement between the aircraft operating locations and the CRFs.[19] As we did for the F-16, we conducted additional analyses to identify how sensitive these alternative KC-135 CRF network strategies were to variations in shuttle cost. The KC-135 CPFH used was $5,370.[20] For the KC-135R, aviation fuel costs were $3,278, or 61 percent of the total CPFH. Because the shuttle

---

[17] The costs presented for the AD/AFRC network and the TF network in Figure 4.9 assume that the UTC-based AMXS is implemented for the AD/AFRC and TF, respectively, independent of a split operations capability.

[18] We assumed an RC drill-position personnel cost of 25 percent of the AD personnel cost of $65,000. For those RC positions that are assumed to be activated in support of steady-state deployed operations, we assumed an additional personnel cost of $65,000, equal to the AD personnel cost.

[19] This shuttle cost is presented only for home-station operations because of the uncertainty associated with deployed operating locations.

[20] As in the F-16 analysis, this figure was based on U.S. Air Force, 2006, Table A4-1.

costs are small relative to the other costs presented in Figure 4.8, the AD/AFRC CRF network alternative would be less expensive than the current system even if the CPFH increased up to a factor of 32 times the $5,370 figure, or, holding all other CPFH components constant, if the price of aviation fuel increased up to a factor of 51 times the $3,278 figure. Similarly, the TF CRF network would be less expensive than the current system even if CPFH increased up to a factor of 27 times the $5,370 figure, or if the price of aviation fuel increased up to a factor of 43 times the $3,278 figure (holding all other CPFH components constant).

The facility costs associated with the establishment of CRFs are also presented for the maintenance network alternative; however, they amount to a small fraction of the total annualized costs. This suggests that, although the facility costs in this analysis are perhaps somewhat underestimated, even if they were understated by a factor of 10 they would not be so large as to have a material effect on the conclusions.

As discussed earlier, the Air Force could also choose to implement an alternative lying between these two endpoints of "enhanced effectiveness" and "increased efficiency" for KC-135 maintenance. For example, it could select a posture that adds a split-operations capability to some, but not all, CA squadrons, if it wished to capture some effectiveness increases while also allowing some reallocation of resources to career fields other than aircraft maintenance.

CHAPTER FIVE

Conclusions

The analyses described in this monograph focused on identifying alternatives for rebalancing the resources invested in MG maintenance with those invested in the CRF maintenance network for the F-16 and KC-135, from a TF perspective—including the AD Air Force, along with the AFRC and ANG. Assuming a required capability level, a tradespace of alternatives was identified whose endpoints range from an "enhanced operational effectiveness" option that increases capability beyond its current level at no additional cost to an "increased efficiencies" option that meets the capability requirement at a significantly reduced cost. For both MDSs, this range of alternatives is made possible by the centralization of non-MG maintenance activities into a small number of CRFs, which allows for significant reductions in maintenance manpower requirements due to economies of scale.

F-16 Results

For the F-16, the endpoints of this tradespace allow the following alternatives:

- the creation of a split-operations capability in AS maintenance at every CC squadron, with no increase in current resources (with the additional 1,900 AS split-operations manpower positions offset by similar backshop reductions achieved through CRF network economies)

- an annual savings of approximately $90 million, with no reduction in backshop capability to support flying operations, if the Air Force believes that its current F-16 maintenance operational capabilities are sufficient and elects to capture the savings associated with backshop centralization efficiencies, not adding a split-operations capability to the CC squadrons.

KC-135 Results

Even though the logistics requirements and operational demands of the KC-135 are quite dissimilar from those of the F-16, this analysis found that it exhibits a similar potential for effectiveness and efficiency gains by using a CRF maintenance network for all non-MG maintenance workloads. Assuming the implementation of AMC's FOL/RMF maintenance concept on KC-135 home-station operations, the analysis identified a tradespace with endpoints that allow the following alternatives:

- the creation of a UTC-based split-operations capability in AS maintenance at every CA squadron, with no increase in current resources (the AS would be increased by 200 manpower positions in the AMXS to support UTC requirements, with an additional 2,400 split-operations positions in the AS, all offset by backshop reductions achieved via CRF network economies)
- an annual savings of approximately $100 million, with no reduction in backshop capability to support flying operations if instead the Air Force elects to capture the savings associated with backshop centralization efficiencies, not adding a split-operations capability to the CA squadrons.

For both the F-16 and the KC-135, our analyses suggest that the potential exists for improvements in operational effectiveness and/or system efficiency, whether the CRF network supports the TF or only the AD and AFRC forces. If the CRF network supports only the AD/AFRC forces, the associated reduction in backshop manpower is large enough to create a split-operations capability at AD and AFRC squadrons without increasing the baseline total maintenance manpower; resources would not be freed to also generate a split-operations capability at ANG squadrons. While the potential savings associated with the increased-efficiency alternative would be larger for the TF network, there is still an economic case for repair network centralization for an AD/AFRC CRF network.

Of course, the Air Force could also decide to implement a solution lying between these efficiency and effectiveness endpoints, for either MDS. For example, it could decide to create a split-operations capability in some, but not all, CC F-16 squadrons. Other alternatives for reducing manpower requirements exist if the Air Force were to vary the deployment burden or RC participation policies discussed earlier.

However, these alternatives address only resource rebalancing within a single MDS; a broader view should also consider options for rebalancing resources across MDS. For example, were it thought that the future security environment will exert much more stress on mobility aircraft than on fighters, a desirable option might be to centralize F-16 backshop maintenance, using some of the attendant manpower reductions to create a split-operations capability for a small number of F-16 squadrons, and transferring the remaining manpower reductions across other MDSs to help create a split-operations capability for mobility aircraft. Furthermore, rebalancing options should also include the reprogramming of resources between maintenance and other career fields, if projections suggest that those fields will be more stressed in the future security environment. Reviews and assessments of OSD guidance, such as the SSSP, could be used to help the Air Force make such discriminations among aircraft and across career fields.

Next Steps

A subsequent research effort within LEA will address alternatives for wing-level maintenance rebalancing for the C-130. While this MDS is similar in many respects to the KC-135, the large number of special variants (e.g., the AC-130 gunship) presents an opportunity to address the logistics implications of support to high-demand, low-density fleets. In addition, this analysis will address the assignment of additional workloads to the CRF maintenance network (e.g., home-station checks) that go beyond what is included in the KC-135 analysis (and thus beyond what is included in AMC's FOL/RMF construct). Research will also progress on the other fundamental objectives, to identify a range of alternatives for the future Air Force logistics enterprise for the consideration of Air Force logistics leaders.

APPENDIX A

Maintenance Manpower Authorizations

Determining F-16 Maintenance Manpower Authorizations

F-16 wing-level maintenance manpower levels were determined using Manpower Programming and Execution System (MPES) data, which were refined via the procedure discussed in this appendix.

RAND's source for UMD manpower authorizations data is the end-of-month MPES data extract, which we obtain from the Air Force MPES Web site maintained by AF/A1MZ (Air Force Community of Practice, 2008). The MPES data are a consolidation of UMDs for all Air Force organizations and locations into a single data table that contains all Air Force manpower requirements across the TF, including both unfunded manpower requirements and funded manpower authorizations. Our analysis includes only funded authorizations and excludes unfunded manpower requirements. Funded authorizations represent the positions in a unit—not the personnel actually assigned. The authorizations are the basis for planning and programming and thus are the most appropriate measure of manpower resources for analytical purposes. A data snapshot from September 30, 2007, was used for the analysis.

F-16 wing-level maintenance is currently organized under a maintenance group,[1] which comprises four squadrons: AMXS, CMS, EMS,

---

[1] In May 2008, Gen Michael Moseley, Chief of Staff of the Air Force, approved PAD 08-01. This directive realigns bomber, rescue, and fighter (including F-16) aircraft maintenance units into their attendant flying squadrons and transfers all their remaining maintenance functions into a new materiel group. This organizational change was scheduled to be com-

and MOS. The "hands-on" maintenance tasks are performed by personnel assigned to the AMXS, CMS, and EMS, which are further divided into work centers, or "shops," according to the maintenance tasks they perform. Table A.1 presents the set of work centers for an F-16 maintenance group. As discussed in the main body of this monograph, this analysis excludes the following maintenance shops: propulsion flight, avionics work centers (sensor/LANTIRN, avionics test stations, electronic warfare), munitions flight, AGE flight, and survival equipment. Excluded shops are italicized in Table A.1.

The MPES data contain all manpower authorizations, including, but not limited to, maintenance manpower. The Organizational Structure Codes (OSCs) were used as identifying factors to segregate the maintenance manpower records from the entire MPES data set. Each work center can be associated with a unique set of one or more OSCs.[2]

Table A.1
F-16 Maintenance Work Centers

Aircraft Maintenance Squadron	Component Maintenance Squadron	Equipment Maintenance Squadron
Crew chiefs	*Propulsion flight*	Aircraft inspection
Specialists	*Sensor/LANTIRN*	Armament flight
Flightline propulsion	*Avionics test stations*	Wheel and tire
Flightline E&E	*Electronic warfare*	*Survival equipment*
Flightline attack control	Pneudraulics	*Munitions flight*
Weapon loaders	Fuels	Structural repair
Weapon maintenance	Egress	NDI
	E&E	Metals technology
		AGE flight

pleted by November 2008. The implementation of PAD 08-01 was placed on hold in June 2008 following Secretary of Defense Robert Gates's recommendation for Gen Norton Schwartz to serve as the 19th Chief of Staff of the Air Force. While our research team provided information to the Air Force team that was tasked with development of the PAD 08-01 reorganization, the AS structure presented in this monograph could be viewed as an alternative for maintenance reorganization, extending beyond the PAD 08-01 realignments but maintaining a separate maintenance organization.

[2] The link between OSC and work center was identified as presented in U.S. Air Force (2003).

By filtering the RAND database on the appropriate set of OSCs, we identified the maintenance subset of the manpower data.

It is possible to tie the shop manpower back to the squadron level using the OSCs and their associated organizational titles (ORGTs). Manpower totals were computed for each of the four squadrons, along with the maintenance group manpower. Within the CMS and EMS, we further identified the manpower associated with the set of shops that were excluded from this analysis, along with the CMS and EMS remainder manpower (subtracting the excluded shops).

The AFRC and ANG data, however, presented an additional complication, because some records reflected full-time personnel authorizations, while others reflected part-time. To differentiate between full- and part-time personnel, we used the Resource Identification Code and Title (RIC), which identifies personnel type. The RIC field typically indicated each record as belonging to one of four main groups: AD officer/enlisted, drill officers/airmen, ANG/AFRC technicians, or nontechnician civilians. The following logic was used to calculate the full- versus part-time authorizations:

$$Full\text{-}time = AD\ officer/enlisted + ANG/AFRC\ technicians$$
$$+ nontechnician\ civilians$$

$$Part\text{-}time = drill\ officers/airmen - ANG/AFRC\ technicians$$

Units that are assigned both the F-16 and another MDS presented another complication, because it can be difficult to identify the manpower positions that reflect F-16 support and exclude those that provide support to the other MDS. For wings that consist of squadrons that support specific MDSs (e.g., Osan AB, with one F-16 squadron and one A/OA-10 squadron), the AMXS personnel for the non–F-16 squadrons could be easily identified and removed from the manpower counts. It is more difficult to identify the F-16–specific personnel in the AMXS for units that do not have such easily separable squadrons (e.g., Nellis AFB). Furthermore, the CMS and EMS are not organized in a flying-squadron-specific manner for any unit. Thus, we included all CMS and EMS personnel for all units in our counts, along with all AMXS personnel for those multiple-MDS units with non–easily separable squadrons. This introduces a slight overestimate in our count of F-16 maintenance manpower.

Once the distinction between full- and part-time positions for the AFRC and ANG data was made and all multiple-MDS issues were addressed (to the best of our ability), we obtained the manpower authorization counts presented in Table A.2.

Determining Current KC-135 Maintenance Manpower Authorizations

KC-135 wing-level maintenance staffing levels were determined using the same September 30, 2007, MPES data extract as was used for the F-16 analysis.

KC-135 wing-level maintenance is currently organized under a maintenance group, which comprises three squadrons: AMXS, MXS, and MOS. The hands-on maintenance tasks are performed by personnel assigned to the AMXS and MXS, which are further divided into work centers, or shops, according to the maintenance tasks they perform.

Table A.2
F-16 Maintenance Personnel Authorization Totals

| | | Maintenance Personnel Authorization | | | | |
| | | ANG | | AFRC | | |
Operation	AD	Part-Time	Full-Time	Part-Time	Full-Time	Total
Maintenance group and MOS	1,954	598	631	82	98	3,363
AMXS	6,147	2,628	1,674	413	281	11,143
EMS	4,537	1,698	1,018	267	159	7,679
CMS	2,661	1,199	1,271	173	194	5,498
Total	15,299	6,123	4,594	935	732	27,683
Propulsion and avionics	*1,516*	*480*	*693*	*68*	*106*	*2,863*
AGE and munitions	*2,539*	*936*	*363*	*176*	*79*	*4,093*
CMS and EMS remainder	*3,143*	*1,481*	*1,233*	*196*	*168*	*6,221*

NOTE: A total of 125 survival equipment positions are included in this count. While this work center has been removed from the maintenance organization and placed into the operations squadron, a very small number of positions remained in the CMS/EMS manpower counts as of our September 30, 2007, data extract.

Table A.3 presents the set of work centers for a KC-135 maintenance group. As discussed in the main body of this monograph, this analysis excludes the survival equipment work center.

To begin the calculations, we limited the MPES data to maintenance organizations for the worldwide set of KC-135 operating locations. Because many of these locations were assigned multiple MDSs, we utilized the Air Force Program Element Code (PEC) as an additional filter to compile a complete KC-135 data set. Approximately six PECs were used to find all KC-135 manpower in the AD, ANG, and AFRC. In instances where the PEC was unambiguous, we were able to omit non–KC-135 manpower from the total count. When such a determination was less clear, we included all associated manpower positions, introducing a slight overestimate in our count of KC-135 maintenance manpower.

Once the data consisted of only KC-135 positions, we used the ORGT and organizational hierarchy data to compute manpower totals for each of the AMXS, MXS, and MOS, along with the maintenance group.

As in the F-16 manpower analysis, we used RIC data to differentiate KC-135 full- and part-time personnel for the AFRC and ANG.

Table A.3
KC-135 Maintenance Work Centers

Aircraft Maintenance Squadron	Maintenance Squadron
Crew chiefs	Structural repair
Specialists	Aero repair
Flightline propulsion	NDI
Flightline hydraulics	Metals technology
Flightline E&E	E&E
Flightline communication/navigation	Hydraulics
Flightline guidance and control	Fuels
	Propulsion
	AGE flight
	Wheel and tire
	Aircraft inspection
	Survival equipment

Once the distinction was made between full- and part-time positions and all multiple-MDS location issues were addressed (again, to the best of our abilities), we obtained the manpower authorization counts presented in Table A.4.

Table A.4
KC-135 Maintenance Personnel Authorization Totals

		Maintenance Personnel Authorization				
		ANG		AFRC		
Operation	AD	Part-Time	Full-Time	Part-Time	Full-Time	Total
Group and MOS	483	400	309	159	191	1,542
AMXS	2,167	758	585	676	436	4,622
MXS	1,427	1,880	1,471	365	430	5,573
Total	4,077	3,038	2,365	1,200	1,057	11,737

Modeling F-16 Maintenance with the Logistics Composite Model

LCOM is a stochastic simulation model most commonly used to determine the manpower requirements associated with direct aircraft maintenance activities. These activities include the preparation of aircraft on the flightline, the repair of planes and aircraft components that experience a failure during flight operations, and the maintenance of airframes that are due for scheduled maintenance.

The key driver in LCOM is the demand for sorties in a preprogrammed flight schedule. When a sortie is required, an aircraft is selected from the available pool. Once the sortie has been flown, LCOM determines whether any parts or aircraft subsystems require maintenance. If a repair is necessary, LCOM simulates the requisite repair networks that will ensure that the aircraft can be returned to flight operations.

LCOM first handles maintenance requests by verifying that the resources needed for the repair are available. These resources include spares, equipment, and the manpower required to effect the repair. If no spares are on hand or if manpower is unavailable, then either the maintenance task will be deferred or resources previously allocated to another task will be rerouted to it.

To summarize, the user provides a level of resources—manpower, equipment, and spares—that will be allocated to generate sorties for a known flight schedule. LCOM simulates the repair actions needed to produce those sorties from an aircraft pool. The LCOM analyst's task is to determine the quantity of resources that will meet mission requirements at a satisfactory level of service.

The RAND team obtained LCOM maintenance network models for the F-16 from the 2nd Manpower Requirements Squadron (2nd MRS) at Langley AFB. The 2nd MRS used these models to analyze manpower requirements for F-16 maintenance squadrons at both Hill AFB and Cannon AFB. The models include flight schedules for the missions F-16/CG pilots commonly support, such as air interdiction, combat air support, and combat air patrols. LRU stock levels in the model represent typical depths in the Block 40 squadrons at Hill AFB and Cannon AFB. With these models, the team was able to replicate the results found in the LCOM reports published by the 2nd MRS (U.S. Air Force, 2003, 2004), and these results became baseline values for this study.

For this analysis, it was also important to isolate the workloads associated with aircraft phase inspections. To do so, we ported the task network found in the ACC model to a database structure, where the phase task network could be readily isolated from nonphase activities. Within the database structure, tasks that were common to both phase and nonphase networks could be readily identified. New phase-specific tasks were created to distinguish them from similar activities occurring outside of phase. For example, an activity common to both networks is the disassembly of the F-16's machine gun. To track the time spent in phase on this activity, a phase-specific gun disassembly task with a maintenance time distribution identical to the original was added to the phase network.

Much as tasks needed to be segregated to study maintenance specific to phase, identifiers for the maintainers themselves were segregated. To size shops specific to aircraft phase support, manpower categories for phase were created to mirror their nonphase counterparts. For example, when a nonphase task called for an NDI technician, work hours were tallied in the "2A7S2" bin, a moniker related to the AFSC for NDI. Similarly, when a phase-related task necessitated an NDI expert, those hours were tallied separately in a bin labeled "2A7S2P" so that the phase NDI shop could be sized independently of its nonphase sister shop.

The RAND analyses also differed from those in the 2nd MRS reports for Cannon AFB and Hill AFB in terms of the squadron sizes of relevance to the research. The remainder of this appendix walks through the general process for tailoring the baseline LCOM model to these additional squadron sizes.

First, the level of spares must be scaled appropriate to the squadron size. Using the baseline model from 2nd MRS, LRU stock levels were changed according to a simple power-law rule:

$$LRU\ multiplier = \sqrt{\frac{squadron}{baseline}}.$$

This is a rule of thumb consistent with common industrial practice. For example, the model of a 48-PAA squadron would have an additional 40 percent of each LRU found in ACC's model of a 24-PAA squadron. Linear scaling often yielded an excess of spares in test runs, whereas the power law shown here provided a depth of LRUs sufficient for meeting sortie-generation targets.

To determine the sorties required of a new squadron size, the baseline flight schedule was scaled linearly. Thus, the LCOM model of a notional 48-PAA squadron would request twice as many sorties as those found in ACC's 24-PAA baseline.

With these inputs in place, LCOM was then run with an arbitrarily large number of personnel. This informs the analyst of the total flying hours and sortie success rate in the absence of manpower constraints.

ACC's published LCOM analyses reveal that their manpower-constrained F-16 maintenance network can provide approximately 90 percent of the sorties that could be generated in the manpower-unconstrained case. This floor was thus used as a limit in F-16 sortie generation in our analyses. In running the LCOM FOL scenarios, the RAND team noted the total number of phases required to support the simulated flying program. To size maintenance shops in a phase-only facility, the team elected to use 95 percent of this phase count as a lower-bound constraint in LCOM models of CRFs.

Another system constraint provided by the ACC studies that informs the analyses in this monograph is a bound on the NMCS rate. These studies state that NMCS must fall within a window of 11 percent ±5 percent. Given that LRU levels were determined by the power law described earlier, the size of manpower shops was the remaining variable that could be adjusted to ensure that NMCS fell into the acceptable rate window.

As an illustration of the overall LCOM modeling process, consider the following notional analysis of a 96-PAA unit. As inputs to LCOM, LRU stock levels would be set to twice that found in a 24-PAA model from ACC, while sortie counts would be four times those found in the 24-PAA analog. Assume that the model, with manpower as an unlimited resource in a sustained wartime scenario, produces a 98 percent sortie success rate and 8 percent NMCS. The analyst would then reduce the manpower, on a shop-by-shop basis and in a highly iterative fashion, ensuring that the sortie success rate never drops below $0.9 \times 0.98 = 88$ percent and maintaining NMCS between 6 percent and 16 percent. The analyst would then note the number of phases generated annually by this squadron—say, 350—and would require that a modeled CRF have sufficient manpower to produce $0.95 \times 350 = 333$ phases per year.

Once shop sizes are determined that meet each of the system's constraints, these manning values need to be scaled upward to account for MAFs. LCOM assumes that each individual in a maintenance shop is 100 percent available in his work shift for each day of the week the shop is open. An MAF is applied to account for the factors that limit an individual's availability in an actual, nonsimulated shop, such as indirect labor (e.g., filing and classroom training), holidays, and work-week limitations (e.g., 40 hours per week in peacetime and 60 hours per week under wartime rules). The 2nd MRS reports provide specifics on MAF computations, showing the peacetime MAF to be 1.038 and the sustained wartime MAF to be 1.461.

These MAFs are linear multipliers for an LCOM shop size. For example, if LCOM determined that the wheel and tire shop in the above example should have five people in a sustained wartime scenario, the actual manning of the shop should be $5 \times 1.461 = 7.3$, which would round upward to eight individuals needed to man the shop to satisfy the maintenance workload and other system constraints.

Analysis of Phase and Periodic Inspection Maintenance Using REMIS

In strategic-level analyses of Air Force aircraft maintenance, LCOM is an effective tool for quantifying and measuring the operational and economic attributes of labor resources. It is an effective tool for providing Air Force leadership with estimates of the operational maintenance tasks and maintenance personnel required to support a desired flying program. Our research plan utilized LCOM as a primary source for determining scale economies in maintenance manpower for wing-level maintenance tasks. A significant fraction of this workload is generated by the scheduled maintenance processes of phase inspections and PEs. Inherent in LCOM is the simulation's estimate for the flow time of the phase/PE process, a number that is important for determining the number and size of CRFs that are required in the repair network. On the basis of the age of the existing F-16 and KC-135 LCOM models (circa 2003 and 1999, respectively) and our preliminary analysis (including numerous site visits to F-16 and KC-135 units), we undertook an examination of phase/PEs beyond the methods incorporated in LCOM. In addition to developing an understanding of actual maintenance practices, we were interested in validating key data parameters derived from the LCOM results. We chose to gather and analyze empirical field data, as recorded in the Air Force's REMIS maintenance data collection system. By using both LCOM and REMIS estimates to analyze and quantify the attributes of the phase/PE process, we gained a deeper understanding of the performance and operational practices currently used by the Air Force.

Our methods and objectives in this analysis consisted of several components. First, we performed a basic assessment of REMIS as a tool and its capabilities for properly capturing the work that occurs during phase/PE maintenance periods. Second, we developed a way of improving the analysis of phase/PE inspection periods by calculating three important metrics associated with the fly-to-fly maintenance periods for the phase/PE process, including the fly-to-fly time, the phase/PE flow time, and the labor hours of a fly-to-fly phase. Third, we analyzed attributes of phase/PE inspections across aircraft types as well as MAJCOMs and components. Finally, we compared the fly-to-fly times to the LCOM estimates for phase/PE inspections.

With the assistance of the REMIS Office, we collected three years (October 2004–October 2007) of maintenance data for the F-16 and KC-135 fleets. The F-16 and KC-135 data samples contained a total of 12,582,791 and 5,797,385 maintenance labor hours, respectively. We considered only on-equipment maintenance actions, because the focus here is maintenance associated with phase/PE inspections.

Aggregate Analysis of REMIS Data

We began by aggregating all on-equipment maintenance actions by Type Maintenance Code (TMC) to weigh the contribution of phase/PE-type maintenance to the total labor on the aircraft. Figure C.1 presents a distribution of on-equipment maintenance labor hours based on TMC for the F-16 and KC-135. For the F-16, unscheduled maintenance accounts for the largest portion, 53.1 percent. Maintenance labeled "Phase" accounts for approximately 10.2 percent of the total on-equipment labor hours. For the KC-135, unscheduled maintenance is again the largest portion at 38.4 percent. PE maintenance (TMC "Phase") represents a significantly high proportion of on-equipment maintenance, 30.4 percent.

These summary statistics show the importance of the phase/PE process as a source of scheduled maintenance labor hours and thus provide baseline motivation for examining it. Although the phase/PE workload as computed here constitutes a high percentage of the total

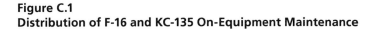

Figure C.1
Distribution of F-16 and KC-135 On-Equipment Maintenance

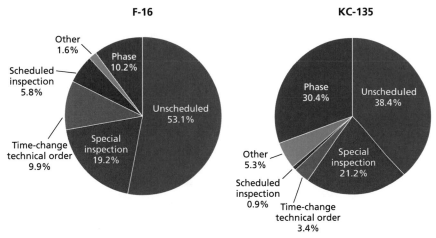

maintenance workload, field discussions indicated to us that the labor hours associated with phase/PE were underrepresented by filtering solely on the TMC designation in REMIS. Many other tasks were being accomplished on the aircraft during the phase/PE period. For example, work on the fuel systems is often performed in the period following the last sortie prior to phase, but before the start of the phase/PE inspection. Similarly, the phase/PE period is frequently used to complete deferred discrepancies that have accumulated on the aircraft. Maintenance actions such as these are often not labeled as "phase" and are therefore not captured in REMIS as phase/PE workloads. However, they are commonly performed as a part of the phase process, when panels are removed from the aircraft and maintainers can easily access subsystems and components.

We were interested in capturing all of the workload that is performed during a phase/PE interval, including the repair requirements identified by the inspections and other scheduled maintenance actions that were synchronized with the phase/PE process. We observed that the maintenance that occurred during phase/PE was often coded with different TMCs—most commonly as unscheduled maintenance. We knew of no existing process for aggregating the data and calculating the contribution of other maintenance actions to the phase/PE process. Therefore, we developed a process to capture all maintenance that occurs in the fly-to-fly period surrounding a phase/PE. We discuss our approach and our findings in the following sections. This approach was developed and verified (by searching through several thousand maintenance records) with the assistance of a career Air Force maintenance officer.

Fly-to-Fly Phase Process

Early in our efforts, we conducted several interviews in AD, ANG, and AFRC F-16 and KC-135 units to learn more about the phase/PE process. The metric "number of days to complete a phase" was consistently mentioned as the primary measurement of performance. However, we found that units have drastically different definitions of the phase/PE process, the tasks included during phase/PE, and the actual time it takes to complete phase/PE. The minimum tasking definition included only phase/PE-related inspections in the "-6" manual. At the other end of the spectrum, some units classify the phase/PE process as including the inspection period along with other scheduled maintenance, time-change technical orders, delayed discrepancies, and preventive maintenance actions. Any attempt to reconcile these differences between units is complicated by the fact that different levels of personnel and personnel shifts are staffed to complete a phase at different bases and MAJCOMs. Therefore, comparing units and their number of days until completion for phase/PE may not be appropriate unless the calculation is normalized. We developed an approach to capture all the additional tasks, the labor hours, the phase/PE flow rate, and the aircraft downtime associated with phase/PE maintenance. These values not only provide more realistic performance metrics of phase/PEs than other approaches do, they allow us to better estimate the flow time of a phase/PE, which impacts the number of facilities required to perform large-scale regionalized or centralized phase/PE operations.

In REMIS, we began by sorting on-equipment maintenance records by each aircraft serial number and the date and time of maintenance action. We used the five-digit WUC to identify phase/PE events. We subsequently found a few dozen instances where a phase/PE WUC identified in REMIS was clearly not a phase/PE event. However, by searching for a larger number of phase/PE-matching WUCs in combination, we were able to filter a significant portion of the data discrepancies. We then sorted aircraft sortie records by aircraft serial number and the date/time of sortie. We then combined the maintenance and sortie records by each aircraft and date/time. The graph on the left side of Figure C.2 shows the daily maintenance hours accrued against a single aircraft over an interval, illustrating how the inclusion of only tasks whose TMC is labeled "Phase" (shaded in dark green) understates the total amount of on-equipment maintenance performed during the fly-to-fly interval. The graph on the right side of the figure illustrates how other on-equipment maintenance (shaded in light green) can be captured during the fly-to-fly interval, using our approach.

Figure C.2
Determining Phase/PE Maintenance and Fly-to-Fly Phase Maintenance Periods

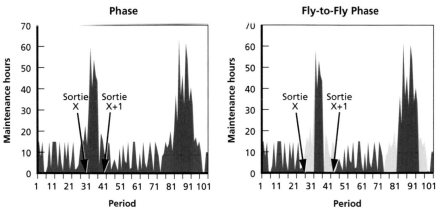

After we categorized all maintenance records into periods of fly-to-fly phase/PE maintenance and other maintenance, we summed all maintenance hours and compared this distribution of hours to our calculations from Figure C.1. Figure C.3 shows that fly-to-fly phase/PE maintenance totals, computed as above, are much higher than the maintenance workloads labeled "Phase" TMCs, for both the F-16 and the KC-135. This is consistent with our assumption that a substantial amount of additional on-equipment maintenance is being performed on the aircraft during the phase/PE period. On-equipment maintenance labor hours that are designated as fly-to-fly phase/PE are now represented by two shades of green, the dark signifying maintenance flagged as "Phase" in the TMC and light green signifying unscheduled maintenance, time-change technical orders, special inspections, and other maintenance that occurred during a fly-to-fly phase/PE. In total, fly-to-fly phase represents 17.6 percent of all scheduled maintenance for the F-16 (with an additional 2.5 percent for engine phase), and fly-to-fly PE represents 45.2 percent of scheduled maintenance for the KC-135.

Figure C.3
Distribution of F-16 and KC-135 On-Equipment Maintenance Using Fly-to-Fly Times

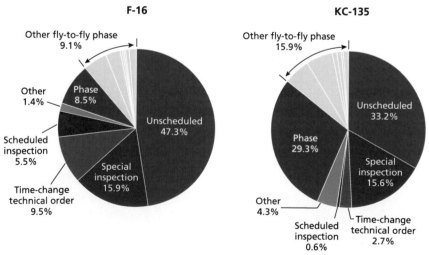

RAND *MG872-C.3*

The adjustment in maintenance hours labeled "Phase" is attributed to maintenance that is associated with other phase/PE maintenance, such as engine phase maintenance.

The aggregate distribution of maintenance hours in Figure C.3 shows that the fly-to-fly phase/PE process is an even greater portion of on-equipment maintenance than was previously understood. However, these aggregate data do not provide insights into the operational performance or metrics of the phase/PE process such as average fly-to-fly phase/PE times, differences between MAJCOM phase/PE flow times, or differences between labor hours by aircraft types (series or blocks).

We performed further analysis on the data and captured three metrics important to fly-to-fly phase/PE events. First, we determined the fly-to-fly phase/PE time, which we define as the number of days between the last sortie that precedes the phase/PE and the first sortie that follows the phase/PE. Second, we calculated all on-equipment labor hours for the fly-to-fly phase/PE event. Third, we developed a technique to estimate the flow time of a phase/PE event. We estimated flow times by identifying every hour during the phase/PE fly-to-fly interval in which aircraft maintenance was performed and then summing these unique hours. We discuss this calculation in further detail in the next section.

A summary of the results for the F-16 is displayed in Table C.1, where the data are organized by F-16 block type. The mean and standard deviation were calculated for each of the three metrics. For the F-16, the labor-hour calculations suggest that maintenance differences exist among block types. Table C.2 aggregates the data by MAJCOM

Table C.1
F-16 Fly-to-Fly Times, Flow Times, and Labor Hours, by Block

Block	Fly-to-Fly Times (days)		Flow Time (days)		Labor Hours	
	Average	Std. Dev.	Average	Std. Dev.	Average	Std. Dev.
25	36.7	25.4	9.6	3.9	730	342
30	44.4	29.1	10.1	4.9	920	482
40	19.6	10.9	7.7	3.0	955	717
42	31.0	22.2	8.5	4.1	691	429
50	20.1	11.7	7.3	2.8	742	366

Table C.2
F-16 Fly-to-Fly Times, Flow Times, and Labor Hours,
by MAJCOM/Component

MAJCOM/Component	Fly-to-Fly Times (days)		Flow Time (days)		Labor Hours	
	Average	Std. Dev.	Average	Std. Dev.	Average	Std. Dev.
ACC	19.5	12.0	7.3	3.0	854	816
AETC	23.5	10.2	7.4	1.8	498	188
ANG	49.9	29.3	11.0	5.1	966	460
AFRC	32.4	16.3	9.3	3.6	872	396
USAFE	20.9	12.7	8.3	3.2	936	937
PACAF	21.7	11.8	7.6	2.8	1,360	1,338
AFMC	29.6	14.4	11.6	4.8	793	793

and component. Figures C.4, C.5, and C.6 are histograms that display the variation of labor hours, fly-to-fly times, and phase flow times across the F-16 blocks.

Some of the variation observed at the block level may be due to the aircraft age and hours flown. However, the differences across block numbers are largely influenced by another factor, namely the fact that certain block numbers are primarily assigned to certain MAJCOMs and components, e.g., most of the Block 30 F-16s are assigned to the ANG. Organizing the data by MAJCOM and component reveals significant differences among the fly-to-fly times of AD, ANG, and AFRC units. The variances in fly-to-fly intervals may be partially explained by the fact that ANG and AFRC units typically operate one daily maintenance shift, while their AD counterparts often perform two-shifts-per-day operations. However, the flow-time metric was designed to minimize the effects of shift scheduling policies on phase/PE production times; thus, these differences in the number of daily shifts should not have a great effect on the phase flow time or on labor hours. The remaining differences in phase flow time and labor hours across MAJCOMs and components result from several factors, including the number of maintenance personnel assigned to phase, variances in the maintenance that is synchronized with or deferred during the fly-to-fly phase period, the policy on addressing all discrepancies during phase,

Figure C.4
Histograms of F-16 Labor Hours

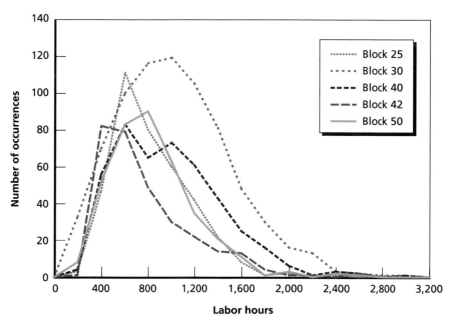

and base-level decisions to maintain the aircraft above and beyond Air Force standards.

Table C.3 displays similar computations for the KC-135, differentiated by MAJCOM or component. As in the F-16 phase calculations, the large variations in KC-135 fly-to-fly time for AD versus those in ANG and AFRC units are influenced by differences in the number of daily shifts assigned to PE, with AD units often performing two shifts per day and ANG and AFRC units typically performing a single daily shift. The remaining differences in phase flow time and labor hours across MAJCOMs and components are again influenced by such factors as the number of maintenance personnel assigned to phase, variances in the maintenance that is synchronized with or deferred during the fly-to-fly phase period, the policy on addressing all discrepancies during PEs, and base-level decisions to maintain the aircraft above and beyond current standards.

Figure C.5
Histograms of F-16 Fly-to-Fly Times

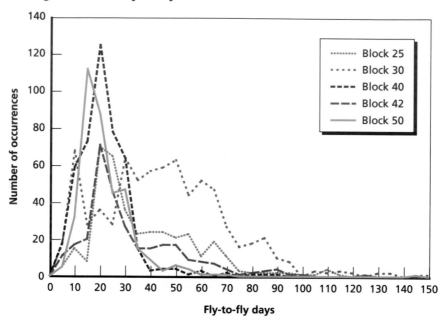

Figure C.6
Histograms of F-16 Flow Times

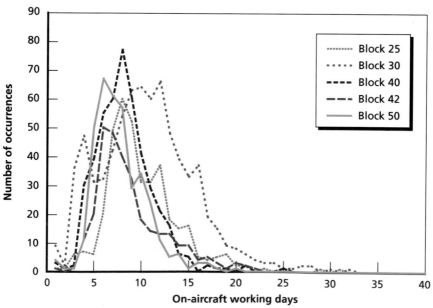

Table C.3
KC-135 Fly-to-Fly Times, Flow Times, and Labor Hours,
by MAJCOM/Component

MAJCOM/ Component	Fly-to-Fly Times (days)		Flow Time (days)		Labor Hours	
	Average	Std. Dev.	Average	Std. Dev.	Average	Std. Dev.
AMC	21.7	12.0	8.5	3.6	1,283	669
AETC	22.5	7.1	8.1	1.8	1,026	267
USAFE	25.7	24.6	10.7	3.3	1,658	513
AFRC	40.5	27.5	12.8	6.6	1,994	1,272
ANG	55.9	24.6	16.0	7.0	2,719	1,409
AFMC	34.7	8.6	13.9	6.4	1,985	1,049
PACAF	43.7	46.7	11.5	5.2	1,153	553

Comparison of LCOM and REMIS

Our approach for estimating a "pure" phase/PE flow time from REMIS data is somewhat flawed, because the flow-time calculation is a function of the number of personnel simultaneously working on the aircraft across the phase/PE process. Operating locations with smaller phase/PE personnel teams may have higher flow times than those with larger teams because of the increased amount of work that can be performed in parallel by larger teams. That said, we believe our flow-time calculation is a significant improvement over fly-to-fly values for the purpose of measuring phase/PE flow times, and to our knowledge, it is the first attempt at quantifying such a phase/PE process flow time. In the LCOM analysis, we derived an F-16 phase flow-time estimate by assigning an arbitrarily large number of personnel to the phase dock, which allows the maximum possible amount of work to be performed in parallel and never delays an aircraft in a queue because of insufficient maintenance manpower. The LCOM-based estimate for F-16 Block 40 phase flow time was 7.5 days. The REMIS-based mean Block 40 phase flow time was 7.7 days, as shown in Table C.1, so these estimates are very consistent. If the LCOM Block 40 estimate is compared with the MAJCOM and component flow times presented in Table C.2, the LCOM flow time estimate of 7.5 days is consistent with the REMIS average for most AD MAJCOMs, with the exception of AFMC (assuming that most of the variation in flow times across blocks is due to MAJCOM-related factors), but the ANG and AFRC REMIS flow times are considerably longer (11.0 and 9.3 days, respectively). As discussed earlier, this is likely due to the number of personnel that are assigned to phase inspection in the ANG and AFRC, along with variances in the amount of maintenance performed during the phase fly-to-fly interval. Overall, we were encouraged by the similarities between the REMIS fly-to-fly phase estimates and the LCOM estimates.

We believe that we have developed an appropriate alternative method of capturing critical maintenance information for use in the LEA and similar studies. Furthermore, our approach and findings suggest additional opportunities, insights, and metrics for the consideration of the maintenance data analysis community in the Air Force. It is beyond the scope of this study to determine exactly what tasks should or should not be included in a phase/PE at a CRF. However, it may be cost effective to assign centralized phase/PE facilities formal responsibility for more tasks than just the minimum inspections and associated repairs. The inclusion of other tasks, such as the workloads captured in this fly-to-fly maintenance analysis, could reduce non–value-added time for flightline maintainers, as well as the total maintenance labor hours associated with the aircraft.

Integer Linear Programming Model

We formulate the facility location problem as an ILP model with an objective function that minimizes the annualized costs of the CRF network. The primary cost drivers for this problem are construction costs for hangars and equipment, transportation costs for shuttling aircraft between operating locations and CRFs, and maintenance personnel costs. Therefore, the decision variables of the model are the number, location, and hangar capacity of opened CRFs; the assignments of maintenance from operating locations to CRFs (it is assumed that these assignments are exclusive, that is, that an operating location has its CRF workload assigned to exactly one CRF); and the personnel required at the CRF.

F-16 Model

We begin by defining the sets used in the decision model. Let J be the full set of candidate CRF locations in the network, with index $j \in J$. Let I be the set of aircraft operating locations, with index $i \in I$. Let K be the set of capacity increments (personnel, facilities), with index k, where $k = 1, 2, \ldots |K|$. A primary assumption in our model is that repair capacity exists in the form of "repair teams." We define a repair team as follows: Assume that the mean flow time for an F-16 phase is eight days, implying that a fully utilized phase dock, operating 24/7, could generate approximately 45 aircraft phases per year. We define a repair team as the capacity required to sustain such a workload level, including all attendant component repair for which the CRF would be responsible. The facilities capacity associated with a given repair team capacity can then be identified in a similar manner.

Repair capacity is defined in such an incremental fashion because of the existence of significant manpower scale economies in CRF operations, as discussed in the main body of this monograph. A simple linear formulation would be unable to capture the manpower savings associated with consolidating larger amounts of work into a single facility. The use of incrementally defined repair capacity allows us to capture these significantly nonlinear effects via the use of piecewise-linear functions. The decision variables can then be defined as follows:

X_{jk} = Boolean decision to assign CRF j incremental facility capacity k (i.e., facility capacity greater than or equal to k)

Y_{ji} = Boolean decision to assign operating location i's CRF workload to CRF j

Z_{jk} = Boolean decision to assign CRF j incremental repair team capacity k (i.e., repair team capacity greater than or equal to k).

Demand at an operating location is dependent on the flying hours assigned to that location. For modeling simplicity, the phase inspection repair times are assumed to be deterministic, and we assume the CRFs operate three shifts per day, seven days per week, 50 weeks per year. The input parameters and cost parameters to the model are defined as follows:

λ_i = annual demand rate at operating location i (associated with a PAA and flying schedule) for CRF maintenance

ρ = total number of phases completed per repair team per year

C_{ji}^T = round-trip shuttle cost for flying an aircraft between i and j

C_{jk}^F = annual amortized facility costs associated with the addition of incremental capacity k at CRF j, including hangar space, shops, and equipment

C_{jk}^P = annual personnel cost associated with the addition of incremental repair team k at CRF j.

The full formulation of the model used to determine the optimal F-16 fixed network follows. The objective function is the sum of the costs, including annual shuttle cost for transporting aircraft to and from the CRF, annualized CRF facility and equipment costs, and annual CRF personnel costs.

$$\text{minimize} : \sum_{j \in J} \sum_{i \in I} C_{ji}^{T} Y_{ji} \lambda_{i} + \sum_{j \in J} \sum_{k \in K} C_{jk}^{F} X_{jk} + \sum_{j \in J} \sum_{k \in K} C_{jk}^{P} Z_{jk}$$

subject to

$$\sum_{j} Y_{ji} = 1, \quad \forall i \in I \tag{D1.1}$$

$$\rho \sum_{k} Z_{jk} \geq \sum_{i} \lambda_{i} Y_{ji} \quad \forall j \in J \tag{D1.2}$$

$$\sum_{k} X_{jk} \geq \sum_{k} Z_{jk} \quad \forall j \in J \tag{D1.3}$$

$$X_{jk} \geq X_{j(k+1)} \quad \forall j \in J, k = 1, 2, \dots, |K| - 1 \tag{D1.4}$$

$$Z_{jk} \geq Z_{j(k+1)} \quad \forall j \in J, k = 1, 2, \dots, |K| - 1 \ . \tag{D1.5}$$

Constraint (D1.1) ensures that every operating location i is assigned to exactly one CRF. Constraint (D1.2) requires the CRF j personnel capacity to be large enough to accommodate its assigned annual demand. Constraint (D1.3) creates a direct relationship between facility capacity and personnel capacity, forcing the facility size at CRF j to be large enough to accommodate the required manpower. Constraints (D1.4) and (D1.5) enforce the piecewise-linear functional relationship, restricting the sequence in which personnel capacity and facility capacity can be purchased. With potential economies of scale, the constraints force the early (and relatively more expensive) increments of capacity to be purchased first.

We note that the X_{jk} and Z_{jk} variables could be combined, which reduces the complexity of the problem. We used separate decision variables to match the data we received as inputs and to increase the readability of the model.

KC-135 Model

A different model formulation was necessary for the KC-135 analysis, because deployed operating locations generally do not perform PE inspections. Rather, aircraft are typically cycled out of the deployed locations, and PE inspections are performed at home-station facilities to avoid the additional deployment burden associated with sending PE personnel and equipment forward into the theater. While KC-135 PE could potentially be performed at a forward-deployed location, this can generally be avoided because of the relatively long KC-135 PE interval.

This creates two important distinctions from the F-16 model. First, deployed KC-135 must be differentiated from home-station aircraft, because home-station aircraft generally breach the 15-month interval between PEs without accumulating 1,500 flying hours, while deployed KC-135 can accumulate 1,500 flying hours in much less time than 15 months. Second, we assumed that if a CONUS CRF is supporting deployed aircraft, it would maintain a peacetime MAF broadly consistent with a standard 40-hour workweek (see Appendix B for more discussion of MAFs). If a deployed CRF were stood up, it would be assumed to be operating under a wartime sustained MAF, broadly consistent with a 60-hour workweek. This suggests that the deployed-manpower requirement per repair team would be less than the home-station-manpower requirement for an equivalent number of repair teams.

We redefine the following sets, parameters, and variables and augment our earlier model formulation as follows:

R = set of capacity increments (personnel, facilities) supporting an increment of deployed aircraft, with index r, where
$r = 0, 1, \ldots, |R|$

X_{jrk} = Boolean decision to assign CRF j incremental facility
capacity necessary to support at least r increments of
deployed aircraft and at least k increments of home-
station aircraft

Z_{jrk} = Boolean decision to assign CRF j incremental repair
team capacity necessary to support at least r increments
of deployed aircraft and at least k increments of home-
station aircraft

C^F = annual amortized facility costs associated with the
addition of an increment of capacity, including hangar
space, shops, and equipment (note that this cost no
longer varies by CRF location or by increment number
and does not differentiate between support of home-
station aircraft and support of deployed aircraft)

C^P_{jrk} = annual personnel cost at CRF j associated with the
addition of the kth incremental repair team supporting
home-station aircraft and the rth incremental repair
team supporting deployed aircraft.

The formulation for the KC-135 model follows. The objective
function is similar to that in the F-16 formulation, with the exception that the costs of both home-station and deployed manpower are
now simultaneously calculated in the model, and a few additional constraints are needed to select the appropriate combinations of deployed
and nondeployed personnel. We formulate the model as

$$\text{minimize}: \sum_{j \in J} \sum_{i \in I} C^T_{ji} Y_{ji} \lambda_i + C^F \sum_{j \in J} \sum_{r \in R} \sum_{k \in K} X_{jrk} + \sum_{j \in J} \sum_{r \in R} \sum_{k \in K} C^P_{jrk} Z_{jrk}$$

subject to

$$\sum_{j} Y_{ji} = 1, \quad \forall i \in I \tag{D2.1}$$

$$\rho \sum_k Z_{jrk} \geq \sum_i \lambda_i Y_{ji} \quad \forall j \in J, r = 0 \tag{D2.2}$$

$$\sum_k X_{jrk} \geq \sum_k Z_{jrk} \quad \forall j \in J, r = 0 \tag{D2.3}$$

$$\sum_k Z_{jrk} \leq 1 \quad \forall j \in J, |r| \geq 1 \tag{D2.4}$$

$$Z_{jr'k} \leq 1 - Z_{jr(k-1)} \quad \forall j \in J, \forall k \in K, |r| \geq 1, r' = 0 \tag{D2.5}$$

$$Z_{jr(k+1)} \leq Z_{jrk} \quad \forall j \in J, \forall k \in K, r = 0 \tag{D2.6}$$

$$Z_{j(r+1)k} \leq Z_{jrk} \quad \forall j \in J, \forall k \in K, \forall r \in R \tag{D2.7}$$

Constraint (D2.1) forces every operating location i to be assigned to a unique CRF. Constraint (D2.2) requires the home-station manpower capacity at a CRF to exceed the assigned home-station demand. Because we assume that the capacity assigned to support of deployed aircraft cannot be used to repair home-station aircraft, we sum this manpower only across $r = 0$. Constraint (D2.3) creates a direct relationship between facility capacity and personnel capacity. Constraint (D2.4) requires that at any CRF j, an assigned level of capacity supporting deployed aircraft can be associated with at most one level of home-station-supporting capacity. Constraints (D2.5), (D2.6), and (D2.7) enforce piecewise-linear functionality that forces the integer program to choose the capacity dedicated to deployed and home-station aircraft in the proper sequence.

Estimating KC-135R Maintenance Manpower Requirements

This appendix describes the process we used to estimate the maintenance manpower requirements for a revised maintenance force structure in which some subset of the KC-135 non-MG maintenance currently performed at bases would be moved to a CRF. We computed shop-by-shop requirements for the CRF and then computed the requirements for the shops (or portions thereof) that remained at the bases. To do this, we generated a table of shop sizes for varying numbers of aircraft sharing a peacetime CRF, as well as a table of the numbers of maintenance personnel that would be required in each shop that remained with the units.

Our estimation process consisted of four steps:

- convert wartime personnel estimates to estimates of the peacetime personnel required to perform the same workload
- perform regression analyses to develop equations for estimating the peacetime personnel required for a CRF in all situations
- use the equations to estimate the requirements for a CRF supporting a specific combination of peacetime and wartime units
- assemble units of various PAA levels from the UTCs and convert the contingency authorizations to equivalent peacetime manpower requirements.

We first describe the background of the data we used in the CRF analysis. We next describe the different data we used to analyze residual-maintenance-personnel requirements at bases of varying sizes.

Finally, two planning tables are presented that specify the maintenance manpower requirements for units of different sizes.

Background

For the F-16 analysis described earlier in this monograph, RAND had to perform new LCOM simulation runs to determine which workloads must remain at the aircrafts' operating locations and which workloads could potentially be performed at a CRF (see Appendix B). In contrast, the KC-135 analysis built on analyses previously conducted by AMC/XPMMS,[1] extending AMC's existing FOL/RMF construct. These analyses separated the aircraft maintenance tasks normally performed at base level into two distinct categories: those whose frequency and operational impact required that they be performed at a deployed FOL in wartime, and those that occurred less frequently (mainly PEs) or that would not immediately affect wartime operations and could therefore be postponed and consolidated with the PEs (i.e., delayed discrepancies). The primary extension in our analysis was the application of this FOL/RMF construct to KC-135 home-station operations.

The AMC manpower analysts who developed the task categories identified individual shops whose operations could be completely remote from the flightline without affecting near-term operations, as shown in Table E.1. Shops that must be retained near the flightline are shown in Table E.2. Shops whose tasks could be split, so that some could be performed remotely without affecting daily operations, although some shop capability would still be required at the FOL, are shown in Table E.3.

The analysts then used LCOM to estimate the maintenance manpower requirements for individual flightline shops and backshops for a wide variety of operational conditions.[2] LCOM is a base-level simu-

---

[1] The source data for this AMC KC-135 analysis and a description of how those data were developed are given in U.S. Air Force (1999).

[2] The nondestructive inspection and AGE shops were not simulated using LCOM in the AMC report. Instead, manpower standards were used to determine the manpower requirements for these shops at the CRF and FOL.

Table E.1
Potential Remote Shops

Backshop	Skill (AFSC)
PE	2A5X1
Engine maintenance	2A6X1
E&E	2A6X6
Hydraulics	2A6X5
Wheel and tire	2A5X1

Table E.2
Shops Required at the FOL

Flightline Shop	Skill (AFSC)
Crew chief	2A5X1
Specialists: communications and navigation	2A4X2
Specialists: guidance and control	2A4X1
Specialists: propulsion	2A6X1
Specialists: hydraulics	2A6X5
Specialists: E&E	2A6X6

Table E.3
Shops That Can Be Partially Removed from the FOL

Backshop	Skill (AFSC)
Aero repair	2A5X1
Metals	2A7X1
Structures	2A7X3
Fuels	2A6X4
Nondestructive inspection	—
AGE	—

lation of aircraft maintenance shops, their workloads, their available maintenance manpower, and the joint effects of workloads and maintenance manpower on achievable OPTEMPO. Manpower analysts use LCOM to construct a mix of skills in shops that have sufficient capacity to meet a target OPTEMPO by adjusting the number of personnel available for operations during each shift.

The KC-135R analyses documented in the AMC/XPMMS report examined the following ranges of operational variables for operations at peacetime bases:

- ASD = 3.5 to 4.5 flying hours per sortie
- UTE = 1.0 to 1.75 hours per aircraft-day
- PAA = 6 to 54 per base.

For each combination of these variables, the analysts constructed KC-135R maintenance shops with sufficient personnel per shift to meet the specified peacetime OPTEMPO.

Next, the analysts used the LCOM model to estimate the maintenance manpower requirements for the same shops in wartime, assuming that only high-priority tasks would be performed at an FOL and that lower-priority tasks would be performed at an RMF distant from the area of operations. The range of wartime cases used to estimate the RMF requirements was more constrained:

- ASD = 4.50 flying hours per sortie
- UTE = 6.75 flying hours per aircraft-day
- PAA = 48 to 96 per base.

It was assumed that the RMF would perform most component repair (excluding wheel and tire repair), as well as the PEs.[3] The cases examined for the FOLs were slightly less constrained:

- ASD = 4.50 flying hours per sortie
- UTE = 4.50 or 6.75 flying hours per aircraft-day
- PAA = 6 to 20 per base.

After running the LCOM simulations, the manpower analysts converted the number of personnel per shift to an estimate of the number that would need to be assigned to each shop, accounting for peacetime or wartime shift length, overtime allowances, leave, and other military duties. They used two factors: shift length (eight hours in peacetime, 12 hours in sustained wartime operations) and a MAF

---

[3] Since 1999, the KC-135 SPO and AMC have extended the PE intervals from a purely isochronal 360 days to either 1,500 flying hours or 15 months, whichever occurs first. They have recently proposed extending the interval further, to 1,800 flying hours or 18 months.

(161.2 hours per month in peacetime, 247.0 hours per month in sustained wartime operations).[4]

Using the FOL/RMF construct, AMC has developed a number of deployment packages for maintenance manpower required to support contingency KC-135R deployments. In these packages, called UTCs, there is an initial maintenance complement for a minimum deployment of four KC-135Rs and additional incremental personnel requirements allowing for the addition of one, two, or four KC-135Rs to the deployed unit. These UTCs provide an independent, operationally tested estimate of the FOL maintenance manpower requirements.

Analytic Approach

Step 1: Convert Estimates of Wartime Maintenance Manpower Requirements to Peacetime Estimates

This conversion requires a few simple calculations. An equation is needed to translate the LCOM wartime requirement estimates into estimates of peacetime work for units of different sizes flying different training and operations programs. The peacetime and wartime LCOM analyses use different manpower-requirements factors (M) and shift (S) assumptions.

When the LCOM is run with three eight hour shifts in peacetime, the basic equation for translating from LCOM requirements into actual personnel requirements is

$$P_p = \frac{D * 8 * L_p}{M_p} \, , \tag{E.1}$$

where

P = number of personnel required

L = LCOM estimate of the number of eight-hour work shifts required to be filled per day (i.e., daily available workers required)

---

[4] When applying these adjustments, the AMC analysts consistently rounded up fractional personnel requirements.

M = MAF

D = number of work days in an average month (30.4375).

For two 12-hour shifts in wartime, the equivalent equation is

$$P_w = \frac{D*12*L_w}{M_w} \cdot \qquad \text{(E.2)}$$

The 8 and 12 in Equations (E.1) and (E.2) reflect the number of work hours available per shift, based on the number of daily work shifts, S, in each case. Thus, the number of manpower positions required to fill one maintenance slot in 24/7 operations in peacetime (R_p) is

$$R_p = \frac{D*24}{M_p} \cdot \qquad \text{(E.3)}$$

The equivalent manpower fulfillment rate for wartime (R_w) is

$$R_w = \frac{D*24}{M_w} \cdot \qquad \text{(E.4)}$$

To convert from one to another, we note that

$$R_p*M_p = D*24 = R_w*M_w \cdot \qquad \text{(E.5)}$$

The Rs are simply scaling factors for the Ps:

$$P_p = \frac{P_w*M_w}{M_p} \cdot \qquad \text{(E.6)}$$

Thus, one can convert the wartime manpower requirement into a peacetime requirement by multiplying the wartime requirement by the ratio of wartime to peacetime MAFs (247.0/161.2, or 1.5323).[5] For our calculation, we used the unadjusted, or "fractional," manpower requirement from the wartime RMF cases. That is, we used the data before it had been rounded up to the next full person. To estimate the equivalent values for the peacetime cases for which we had no fractional requirement estimate, we subtracted 0.5 from the rounded-up value to reduce the bias caused by rounding.

Note our implicit assumption that the numbers of personnel required on different shifts do not differ widely in the LCOM estimates. That assumption is probably valid for the 24/7 wartime KC-135 operations, but it may be less so for operations that have a large day/night or weekday/weekend differential.

Step 2: Develop Equations for Estimating Peacetime Maintenance Manpower for the CRF

After converting the RMF maintenance manpower data for each shop to peacetime equivalents, we combined the (adjusted) RMF manpower requirements with the peacetime base manpower requirements in a stepwise forward regression. We began the regression analysis with the simple assumption that the main contributor to the manpower requirement would be flying hours, but the workload might differ somewhat for the RMF, especially for shops whose workload was split (see Table E.3). The initial regression equation was

$$M = a + b * F + c * C + d * F * C ,$$
(E.7)

where

M = (fractional) manpower requirement estimate
a, b, c, d = constants to be determined
F = daily total flying hours for the unit (i.e., UTE × PAA)
C = an indicator variable (0 or 1) indicating whether the data reflect consolidated (i.e., RMF) operations.

---

[5] The authors wish to thank Adam Resnick for identifying a streamlined exposition of this derivation.

Consider, for example, the data for the aero repair shop, shown in Table E.4. The numbers in the first column indicate whether or not the case was from a consolidated shop; the second column presents the product of the indicator variable and the average daily flying hours supported by the shop; the third column shows the average daily flying hours in each case; and the fourth shows the peacetime fractional manpower requirement.

Observe that the nonconsolidated shop requires 24.5 personnel to meet the workload for 94.5 flying hours, whereas a much smaller CRF shop (19.5 personnel) can meet the consolidated workload for 324 flying hours. This is a result of the CRF concept, in which a substantial portion of the aero repair shop's workload is handled at the flightline in direct support of operations under the CRF concept.

Table E.4
Aero Repair Shop Data for CRF Analysis

Consolidated	Consolidated Variable × Flying Hours	Daily Flying Hours	Adjusted Manpower
0.00	0.00	12.0	16.5
0.00	0.00	22.5	16.5
0.00	0.00	24.0	16.5
0.00	0.00	30.0	16.5
0.00	0.00	31.5	16.5
0.00	0.00	31.5	16.5
0.00	0.00	36.0	16.5
0.00	0.00	42.0	19.5
0.00	0.00	48.0	17.5
0.00	0.00	54.0	19.5
0.00	0.00	54.0	20.5
0.00	0.00	72.0	20.5
0.00	0.00	72.0	20.5
0.00	0.00	94.5	24.5
0.00	0.00	94.5	24.5
1.00	648.00	648.0	27.7
1.00	567.00	567.0	25.1
1.00	486.00	486.0	22.5
1.00	405.00	405.0	22.1
1.00	324.00	324.0	19.5

SOURCE: U.S. Air Force, 1999.

The initial regression analysis yielded the results shown in Table E.5. The P-value indicates the probability that the independent variable is unrelated to the shop's manpower requirement. In this case, there is a 25 percent chance that the "consolidate" variable is not statistically significant, but only a very small chance that the other variables and the intercept are not significantly related at p = 0.05.[6]

We next removed any variables that were not significant at the p = 0.05 level. The regression values of the remaining variables were slightly revised, as shown in Table E.6.

Table E.5
Initial Regression Results for the Aero Repair Shop Fractional Manpower Requirement

Variable	Coefficient	Standard Error	t Stat	P-value
Intercept	13.772346	0.50944825	27.0338467	8.7997E–15
Consolidate	–2.0918723	1.83256194	–1.1415016	0.27045829
Consolidate × flying hours/day	–0.0815887	0.01009218	–8.0843447	4.8387E–07
Flying hours/day	0.10565736	0.00945639	11.1731224	5.7468E–09

NOTE: R^2 = 0.94, p = 3.5E–10.

Table E.6
Revised Regression Results for the Aero Repair Shop Fractional Manpower Requirement After Removing the "Consolidate" Indicator

Variable	Coefficient	Standard Error	t Stat	P-value
Intercept	13.6106802	0.4937089	27.5682296	1.4956E–15
Consolidate × flying hours/day	–0.0880194	0.00844779	–10.419219	8.4525E–09
Flying hours/day	0.10832547	0.00924427	11.7181199	1.4463E–09

NOTE: R^2 = 0.94, p = 4.9E–11.

---

[6] The intercept can be viewed as the minimum manpower requirement for a shop. The probability threshold of 0.05 is arbitrary, but it is intended to assure that only factors that are likely to affect the dependent variable are included. Not only does this increase the explanatory value of the resulting equation, it improves the equation's validity and reliability for points not included in the original data set. An F-test is used on the overall equation, and a t-test is used on the individual parameters in the equation.

We then conducted an exploratory forward regression analysis for each shop to determine whether any of the other variables might help explain the variation in maintenance manpower requirements. We also examined whether ASD, PAA, or UTE could significantly enhance the accuracy of the CRF equation for each shop. For the aero repair shop, none of those factors significantly improved the equation's ability to replicate the available data. In the interest of brevity, we present only the results for adding one variable, ASD. As shown in Table E.7, ASD's P-value exceeds 0.05, suggesting that it is not significant.

We used the coefficients in Table E.6 to estimate the aero repair shop maintenance manpower requirement and found the best equation for that shop to be

$$M_{AR} = 13.6107 + 0.1083 * F - 0.0880 * C * F \ . \qquad (E.8)$$

That is, the aero repair shop at a standard peacetime base should require a minimum of 13.6 personnel plus the number obtained by multiplying the daily flying hours at the base by 0.1083. In contrast, the same shop at a consolidated maintenance facility would require the same minimum number, but only $0.1083 - 0.0880$, or 0.0203, additional personnel per flying hour at the bases it supports. The size of the consolidated shop is far less sensitive to flying hours because it needs to dispatch personnel to its own flightline only occasionally, in contrast to a standard base, where a considerable amount of aero repair shop work originates at the flightline. Thus, we would expect the aero repair shop manpower requirement at a CRF to be relatively insensitive to changes in the size or OPTEMPO of the force the shop supports.

Table E.7
Effect of Including ASD on the Accuracy of the Aero Repair Shop Equation

Variable	Coefficient	Standard Error	t Stat	P-value
Intercept	16.4869778	2.32038608	7.10527356	2.4928E–06
Consolidate × flying hours/day	−0.0863593	0.00840361	−10.276452	1.8771E–08
Flying hours/day	0.10735575	0.00911574	11.7769681	2.6988E–09
ASD	−0.7178734	0.56632908	−1.2675905	0.22308217

NOTE: $R^2 = 0.94$, p = 3.0E–10.

Figure E.1 presents a graphical representation of this regression analysis. The solid lines are the aero repair shop's personnel requirements as computed by Equation (E.8), while the individual points on the graph are the original LCOM modelers' findings. The curves are shown separately, because the requirements for smaller flying programs reflect bases that also provide flightline support, while the larger numbers are for CRF operations.

This procedure was applied to all shops, including the flightline shops, because the CRF must also conduct limited flightline operations to receive and generate aircraft when their PEs are due.[7] The results for all shops are summarized in Table E.8. Some of the shop names are duplicates, because the first six shops that conduct maintenance at the

Figure E.1
Comparison of Aero Repair Shop Personnel Requirements from Equation (E.8) with Original LCOM Results

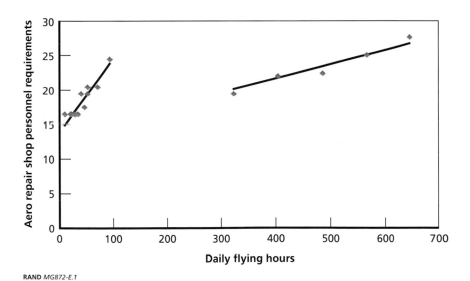

RAND *MG872-E.1*

---

[7] As discussed above, the nondestructive inspection and AGE shops were not simulated in the KC-135 LCOM model. Instead, their manpower was determined using manpower standards. We used the same manpower standards when making our determination of AGE and nondestructive inspection requirements at the CRF and FOL.

Table E.8
CRF Regression Equation Coefficients (Fractional Shop Manpower)

AFSC	Shop Name	Constant	PAA	UTE	Flying Hours/Day	Consolidate	Consolidate × Flying Hours/Day
2A5X1	Crew chiefs	12.4277	1.5*		1.10487		-0.1634891
2A4X2	Communication/navigation	3.9942	0.16513		0.19043		-0.3278
2A4X1	Guidance and control	3.8131			0.33918		
2A6X1	Propulsion	4.0873			0.21661	5.005	-0.2130103
2A6X5	Hydraulics	0.3941			0.492	4.1522	-0.48661
2A6X6	E&E	1.3266			0.42956	3.21896	-0.42596
2A5X1	PE	17.3158	0.26137	0.83941	0.05959		-0.03697
2A6X1	Engines	4.2030			0.19131		-0.16381
2A5X1	Aero repair	13.6106			0.10833		-0.08802
2A5X1	Wheel and tires	3.2577			0.07647	3.5397	-0.07647
2A7X1	Metals	4.7465			0.08897		-0.03298
2A7X3	Structures	10.0467	0.18887	1.40760	0.16445		-0.09026
2A6X6	E&E	5.2564			0.03368		-0.00993
2A6X4	Fuels	3.7868		2.96636	0.40307	1.7931	-0.35000
2A6X5	Hydraulics	3.9548			0.09154		-0.07735

NOTE: The 1.5 factor for PAA covers flying crew chiefs, who were not included in the original LCOM analysis.

flightline use personnel trained in the same skills as those of personnel at the backshops (i.e., the flightline and backshop personnel have an identical AFSC).

The main significant factors for all shops were flying hours per day and the interaction between flying hours and consolidated repair, the latter reflecting the effects of backshop support to flightline operations. Such factors as PAA, UTE, and even simple consolidation were rarely significant at the $p = 0.05$ level.

Some shops have much smaller minimum sizes than others. For example, the flightline hydraulics shop requires less than a whole person per shift because a single hydraulics specialist can perform most flightline work with the support of a crew chief. Further, that skill is required so infrequently that hydraulics specialists can be placed on call or can even accomplish their normal workloads during a single day shift. In contrast, the aero repair shop needs a team of specialists on each shift, and its work requirements occur frequently enough to warrant keeping a full team on duty at all times, even for the smallest unit.

Some shops are more sensitive than others to variations in the flying program. Crew chiefs perform an enormous amount of work to support every sortie, especially on aircraft as large as the KC-135. In contrast, the E&E backshop not only has a smaller workload, it also includes some scheduled inspections unrelated to the actual flying program. Thus, every additional flying hour per day generates a requirement for an additional crew chief, but nearly 30 additional flying hours per day would be needed to justify an additional E&E specialist.

Step 3: Use the CRF Equations to Estimate Shop and Total Manpower Requirements

We constructed a spreadsheet that used the peacetime coefficients in Table E.8 for a CRF serving various combinations of peacetime and contingency units. First, we considered a series of cases in which the CRF supported only aircraft operating in contingency operations.

We used the peacetime equations to estimate the peacetime maintenance manpower requirements for each shop, then used the inverse of Equation (E.6) to convert these requirements into equivalent wartime requirements. In each case, we computed the maintenance manpower requirements for each shop, rounded up to the next integer, then added the requirements across all shops to estimate the total manpower requirements.[8]

For the pure-contingency cases and all other cases in which contingency operations were included, we assumed the following operational parameters:

- UTE = 6.75 flying hours per aircraft-day
- ASD = 4.50 flying hours per sortie
- MAF = 247 hours per person per month
- PE interval = 15 months or 1,500 flying hours[9]
- Contingency PAA supported by the CRF = 0 to 96,[10] in increments of 16.

We then estimated how the requirements would change if the CRF also supported some peacetime operations, using the same parameters we used for the wartime operations, with the following exceptions:

- UTE = 1.75 flying hours per aircraft-day
- MAF = 161.2 hours per person per month

---

[8] Estimates of officer and senior supervisory enlisted personnel were based on the number of personnel they supervised and their office (squadron or flight), using standard manpower policies. For example, every squadron requires a minimum of one officer, and standard planning factors add an additional officer for every 300 personnel in the squadron.

[9] In contingency operations, the 1,500-flying-hour rule is breached prior to the 15-month isochronal maximum, effectively reducing the interval between PEs to 222 days, or 7.3 months.

[10] The PAA upper limit for an accurate equation was constrained by the range of data available for analysis—the largest base in the original LCOM study had 96 PAA. We assumed that larger contingency deployments would require a CRF whose size was proportional to the 96-PAA CRF, in effect assuming that, above that point, there was no further economy of scale. Future LCOM analyses could examine the potential for further efficiencies in larger CRFs; however, such efficiencies would likely be smaller than those for smaller units.

- Peacetime PAA supported by the CRF = 0 to 256,[11] in increments of 32.

Finally, we ran all combinations of peacetime and wartime force sizes, applying a proportional assumption to shops whose workloads would exceed the equivalent workload for a 96-PAA contingency force operating at 6.75 UTE.

Step 4: Estimate Peacetime FOL Maintenance Requirements from UTC Data

Finally, we computed the manpower requirements for the shops remaining at the units. We were able to use maintenance manpower deployment requirements that have been developed to support continuing KC-135E/R deployments in support of Operation Enduring Freedom; Operation Iraqi Freedom; and other, smaller deployments worldwide. These deployments reflect a concept of operations that assumes the implementation of the FOL/RMF construct.

The analysis of contingency operations in the original AMC/XPMMS report used the same concept of operations, but it was conducted more than a decade ago (in 1995), so we were concerned that it might not reflect changes that have been introduced over the intervening period. Therefore, we used UTC data to construct both wartime and peacetime FOL manpower requirements for various PAA levels instead of the AMC LCOM report.

We used UTCs HFKLR, HFK4R, HFK2R, and HFK1R to construct contingency maintenance manpower requirements for KC-135R units ranging in size from four PAA to 16 PAA by adding various combinations of the augmenting UTCs (HFK4R, HFK2R, and HFK1R) to the initial deployment package (HFKLR). We then converted these manpower requirements to peacetime equivalents by again applying the reverse transformation from Equation (E.6). To complete the transformation, we adjusted the fractional manpower requirement for each

---

[11] We selected an upper limit of 256 for applying the equations in peacetime because a peacetime PAA level of 270 would generate approximately the same workload as the contingency PAA level of 96. Again, we assumed that CRFs serving larger forces would require shops whose sizes were proportional to the 270-PAA shop size.

level by assuming that the UTC was intended to support a UTE of 4.5 per day and adjusting the peacetime requirement for flightline personnel using the flying-hours-per-day factor from Table E.8 to reduce the daily flying hours to the equivalent of a 1.75 UTE.

Maintenance Manpower Requirement Estimates

Table E.9 presents the results of our total CRF maintenance manpower estimates for various combinations of peacetime and contingency operations. The values in the gray area are for cases in which the combined workload from the peacetime and contingency operations would exceed the equivalent CRF workload of the largest unit in the AMC report (96 PAA); such cases are considered to be proportional to the 96-PAA-equivalent workload. Cases with more than 544 aircraft in contingency and peacetime operations were not computed. Cases with more than 320 peacetime aircraft were computed, but, to simplify the table, they are not shown.

Table E.10 presents the results for peacetime FOL requirements calculated using UTC data.

Table E.9
Total CRF Maintenance Manpower Requirements

Contin-gency PAA	Peacetime PAA										
	0	32	64	96	128	160	192	224	256	288	320
0	0	141	173	205	234	267	298	328	362	401	445
16	127	190	219	246	277	308	341	375	419	463	507
32	154	234	261	291	320	352	393	437	481	525	569
48	182	278	304	334	362	411	455	498	542	586	631
64	207	319	346	373	428	472	516	560	604	649	693
80	235	359	402	446	490	534	578	623	667	711	756
96	260	420	464	508	552	596	641	684	728	773	817
112	302	482	526	569	614	658	702	746	791	835	879
128	345	543	587	632	676	720	765	809	853	897	941
144	387	605	650	694	738	783	827	870	914	958	1,003
160	430	668	712	755	800	844	888	932	976	1,021	1,065
176	473	729	774	818	862	906	950	995	1,039	1,083	1,127
192	515	792	836	880	924	968	1,013	1,056	1,100	1,144	1,188
208	558	854	898	941	985	1030	1,074	1,118	1,162	1,206	1,251
224	601	915	959	1,004	1,048	1,092	1,136	1,180	1,225	1,269	1,313
240	643	977	1,022	1,066	1,110	1,154	1,198	1,242	1,286	1,330	
256	686	1,040	1,084	1,127	1,171	1,215	1,260	1,304	1,348	1,393	
272	729	1,101	1,145	1,189	1,234	1,278	1,322	1,367	1,411		
288	771	1,163	1,207	1,252	1,296	1,340	1,385	1,428	1,472		
304	813	1,226	1,270	1,313	1,357	1,402	1,446	1,490			
320	857	1,287	1,331	1,376	1,420	1,464	1,508	1,552			
336	899	1,349	1,394	1,438	1,482	1,526	1,570				
352	942	1,412	1,456	1,499	1,543	1,587	1,631				
368	985	1,473	1,517	1,561	1,605	1,649					
384	1,027	1,535	1,579	1,623	1,668	1,712					
400	1,069	1,597	1,641	1,685	1,729						
416	1,112	1,658	1,703	1,748	1,792						
432	1,155	1,721	1,766	1,810							
448	1,198	1,784	1,828	1,871							
464	1,241	1,845	1,889								
480	1,283	1,907	1,951								
496	1,325	1,969									
512	1,368	2,030									
528	1,410										
544	1,454										

Table E.10
Maintenance Manpower Requirements for the Home-Station FOL Concept

	Manpower Authorization												
Shop	4 PAA	5 PAA	6 PAA	7 PAA	8 PAA	9 PAA	10 PAA	11 PAA	12 PAA	13 PAA	14 PAA	15 PAA	16 PAA
MXG/MXOS	11	11	11	11	11	11	11	11	11	11	11	11	11
AMXS supervision	14	14	17	17	17	17	20	20	20	20	23	23	23
Crew chief	24	29	34	38	43	48	53	58	63	68	73	77	82
E&E	6	7	9	11	12	14	15	17	18	20	21	23	24
Hydraulics	6	7	9	10	12	14	15	17	18	20	21	23	24
Propulsion	6	7	9	11	12	14	15	17	18	20	21	23	24
Communication/navigation	7	9	10	12	11	13	14	16	16	17	19	20	20
Guidance and control	7	9	10	12	12	13	15	16	16	18	19	21	21
Structural	4	3	3	3	4	4	4	4	5	5	5	5	6
Aero repair	8	10	8	9	9	10	9	10	10	11	9	11	10
Nondestructive inspection	1	1	1	1	1	1	1	1	1	1	1	1	1
Metals	3	3	3	3	3	3	3	3	3	3	3	3	3
Fuels	4	3	4	3	4	3	4	3	4	3	4	3	4
AGE	9	9	11	11	12	12	14	14	15	15	17	17	18
Total	110	122	139	152	163	177	193	207	218	232	247	261	271

Estimating CRF Component Repair Pipeline Effects

Removing CRF workload from an aircraft's operating location and assigning it to a network facility requires that failed aircraft components be transported between the operating location and the CRF. An inventory of spare components would also be required to support the delay time interval between a component's failure at the aircraft operating location and the receipt of a serviceable replacement from the CRF. For this analysis, we focused on the set of exchangeable aircraft components appearing in both the current RSP for any F-16 unit and the RAND March 2006 capture from the D200 RDB data system. Across all F-16 series and block numbers, this intersection comprises a set of 350 unique NIINs.[1]

The effect on transportation and inventory of removing CRF work centers from the aircraft operating location is limited to that fraction of component failures that were previously repaired at the on-

---

[1] There are 6,372 unique NIINs across all F-16 units' RSPs. However, 89 percent of them have an ERRC code of XB3, indicating that they are expendable, not reparable; such items would generally be disposed of upon failure and would thus not enter into any repair pipeline. Four percent of the RSP NIINs have ERRC code XF3 (authorized for repair at the field level and are generally condemned when the field level cannot return them to serviceable condition), and 7 percent have ERRC code XD2 (authorized for repair at the depot). Our D200 data set for the F-16 contains 1,870 unique NIINs, of which 1,830 have ERRC code XD2, four have ERRC code XF3, and 36 have ERRC code XB3. Of the 350 NIINs included in this analysis, 341 have ERRC code XD2, two have ERRC code XF3, and seven have ERRC code XB3. The lack of XF3 components might understate the effect of backshop centralization, since this would generate a new pipeline requirement for such items.

site backshops.[2] We used two key data elements from the D200 data system to estimate the number of such component failures. The organizational and intermediate maintenance demand rate (OIMDR) identifies the mean number of component failures per flying hour. The BNRTS value identifies the fraction of OIMDR failures that, based on historical analysis, exceed the capability of the on-site backshops to repair. One minus the BNRTS is thus equal to the average fraction of OIMDR failures that are repaired on-site at the base.

If we assume a number of aircraft flying hours to be supported over an interval, we can then compute the expected number of component failures over that interval that would require transportation to an off-site CRF backshop as follows:

$$Flying\ hours \times OIMDR \times (1 - BNRTS).$$

Because this analysis is limited to a subset of the current F-16 backshops, we next needed to identify the backshops responsible for the repair of each NIIN. We used the Air Force Discoverer and Discoverer-Plus data systems to help make this identification (U.S. Air Force, 2008a). A query of this system's repair cycle data table for our set of components associated a stock record account number (SRAN) and NIIN pair with the Organizational Code (ORG) and Shop Code (SHOP) values to arrive at a shop categorization for each NIIN.

Two other Discoverer data tables (Delivery Designation and Organization Cost Center) contain additional information for a given SRAN, ORG, and SHOP and therefore frequently provided hints about whether a given ORG was related to the flightline or to a backshop. Some of these tables provided the shop organizational structure (OSTR), a field that we could decipher using manpower authorization data that provided a description of the OSTR for a unit of corresponding organization number, kind, and type—provided these could be inferred from the SRAN-ORG table (U.S. Air Force, 2008b).

---

[2] The assumption that the transportation of items between the operating unit and the depot would not be affected under this option would be valid as long as the remaining organizational-level maintenance would be able to identify those items that require depot maintenance.

As we analyzed information from these complementary data tables, along with WUCs, federal supply classes and groups, and national stock number (NSN) nomenclature, patterns began to appear that helped us decode many SHOP fields. Although the SHOP field codes are peculiar to each base, there were many repeating patterns, such as the SHOP value being ES for engine shop and MS for metal shop. However, we needed to resolve many apparent conflicting conventions in defining these codes. To do this, we used all other information available about the NIIN. In the end, we arrived at a classification of shops and an identification of which shop repaired a given NIIN or NSN.

We next computed the expected number of component failures associated with each backshop. Assuming a notional home-station monthly flying schedule of 27 flying hours per PAA for the set of NIINs under consideration and counting only those failures that would currently be repaired on-site, we would expect to observe a daily fleetwide average of 74.9 component failures; however, because 35.4 of these failures would be associated with the shops that have been excluded from this F-16 CRF analysis (i.e., JEIM, electronic warfare, LANTIRN, and avionics backshops, accounting for 222 of our total 350 NIINs), only 39.5 of these daily failures were relevant to this analysis.

To estimate the transport costs associated with the use of CRFs in support of home-station operations (including operations for permanently assigned PACAF and USAFE forces), we assumed that all failed components would be shipped using FedEx Small Package Express two-day rates for U.S. domestic shipments. Note that we are not endorsing FedEx here, merely utilizing their cost structure in an attempt to estimate the shipping costs since FedEx is commonly used for shipping such parts. These rates are quoted on a per-pound basis, so it was necessary to obtain weight information for the set of NIINs under consideration. This information was obtained from the D035T database.

Focusing solely on those workloads that were formerly performed within the backshops for the limited set of work centers under consideration, the expected annual F-16 fleetwide transportation cost in support of home-station operations is approximately $700,000.

An inventory requirement can be similarly computed. An additive inventory requirement would be necessary to support the new transportation segments introduced by the removal of component repair workloads to CRFs. The two-day transport time assumed above, in each direction, results in a requirement for four days' worth of pipeline inventory. We estimated this inventory cost utilizing component acquisition cost data from the D200 data system. If we assume that a separate inventory requirement is computed to support each of the permanently assigned USAFE, PACAF, and CONUS F-16 fleets, operating at a notional flying schedule of 27 flying hours per month, a total one-time inventory investment of $4.8 million would be required to support home-station operations.

Because acquisition of this inventory is a one-time additional investment, the cost could be amortized across the expected duration of F-16 CRF operations. Considering an amortization interval as short as five years produces an annualized inventory requirement cost of less than $1 million. Further, a transportation pipeline and inventory requirement would not necessarily be created for every unit, e.g., if the CRF were be located at an existing F-16 operating location.

We performed similar transportation and inventory pipeline computations for the KC-135. Focusing, as before, on the set of aircraft components appearing in both the current RSP for any KC-135 unit and the RAND March 2006 capture from the D200 RDB data system, we identified a set of 298 unique NIINs, across all KC-135 series.

Assuming a notional home-station daily flying schedule of 1.75 flying hours per PAA (for both CONUS-based aircraft and permanently assigned PACAF and USAFE forces), for the set of NIINs under consideration and counting only those failures that would currently be repaired on-site, we would expect to observe a daily fleetwide average of 73.0 component failures. The expected annual fleetwide transportation cost associated with this notional home-station scenario is $2.4 million, assuming, as was the case for the F-16 analysis, that all failed components are shipped using the cost structure associated with FedEx Small Package Express two-day rates for U.S. domestic shipments.

An additive inventory requirement would also be necessary to support the new transportation segments introduced by the CRF. The two-day transport time assumed above, in each direction, for CONUS aircraft, results in a requirement for four days' worth of pipeline inventory. Assume that permanently assigned OCONUS aircraft would be supported from this same inventory pool, with a 14-day transport time to CONUS, in each direction,[3] generating a 28-day pipeline requirement for OCONUS units. If we assume that a single inventory requirement is computed to support the worldwide KC-135 fleet, operating at the notional flying schedule of 1.75 flying hours per day, a total one-time inventory investment of $6.8 million would be required to support home-station operations.

Because acquisition of this inventory is a one-time additional investment, the cost could be amortized across the expected duration of KC-135 CRF operations. Considering an amortization interval as short as seven years produces an annualized inventory requirement cost of less than $1 million. Further, a transportation pipeline and inventory requirement would not necessarily be created for every unit, e.g., if the CRF were to be located at an existing KC-135 operating location.

---

[3] We performed an analysis of Military Aircraft Issue Priority Group 1 (IPG1) shipments for the first ten months of 2007, utilizing the RAND maintained Strategic Distribution Database, which aggregates defense-related pallet movement data feeds, including the AMC GATES database. This analysis suggested an average travel time of 14 days from CONUS to either EUCOM or PACOM.

References

Air Force Community of Practice, "MPES Share" Web site. As of July 30, 2008:
https://wwwd.my.af.mil/afknprod/ASPs/CoP/EntryCoP,asp?Filter=OO-DP-AF-94
(not available to the general public).

Historical Air Force Construction Cost Handbook, Tyndall AFB, Fla.: Directorate of
Technical Support, Air Force Civil Engineer Support Agency, February 2004.

Hoehn, Andrew R., Adam Grissom, David A. Ochmanek, David A. Shlapak,
and Alan J. Vick, *A New Division of Labor: Meeting America's Security Challenges
Beyond Iraq*, Santa Monica, Calif.: RAND Corporation, MG-499-AF, 2007. As of
May 12, 2009:
http://www.rand.org/pubs/monographs/MG499/

International Air Transport Association, "Jet Fuel Price Monitor," updated weekly.
Data used in the analysis appeared on June 27, 2008. As of May 29, 2009:
http://www.iata.org/whatwedo/economics/fuel_monitor/index.htm

McGarvey, Ronald G., James M. Masters, Louis Luangkesorn, Stephen Sheehy,
John G. Drew, Robert Kerchner, Ben Van Roo, and Charles Robert Roll, Jr.,
*Supporting Air and Space Expeditionary Forces: Analysis of CONUS Centralized
Intermediate Repair Facilities*, Santa Monica, Calif.: MG-418-AF, 2008. As of May
12, 2009:
http://www.rand.org/pubs/monographs/MG418/

National Defense Authorization Act for Fiscal Year 2008. As of May 12, 2009:
http://www.govtrack.us/congress/bill.xpd?bill=h110-4986

Ronald W. Reagan National Defense Authorization Act for Fiscal Year 2005,
Public Law 108-325, November 1, 2004. As of May 12, 2009:
http://www.govtrack.us/congress/bill.xpd?bill=h108-4200

U.S. Air Force, *KC-135 Logistics Composite Model (LCOM) Final Report, Peacetime
and Wartime, Peacetime Update*, Scott AFB, Ill.: HQ Air Mobility Command/
XPMMS, May 1, 1999.

———, *USAF War and Mobilization Plan*, Volume 5, *Basic Planning Factors and
Data*, Headquarters U.S. Air Force/XOPW, War and Mobilization Plans Division,
Directorate for EAF Implementation, 2000 (not available to the general public).

————, *Statement of F-16 Block 40 Aircraft Maintenance and Munitions Manpower*, Langley AFB, Va.: Headquarters Air Combat Command/XPM, August 2003.

————, *Statement of F-16 Block 30/40/50 Aircraft Maintenance and Munitions Manpower Cannon AFB, NM*, Langley AFB, Va.: Headquarters Air Combat Command/XPM, February 2004.

————, *Expeditionary Logistics for the 21st Century*, Washington, D.C., April 1, 2005.

————, *US Air Force Cost and Planning Factors, BY 2008 Logistics Cost Factors*, Air Force Instruction 65-503, Annex 4-1, Washington, D.C.: Air Force Cost Analysis Agency (AFCAA), February 22, 2006.

————, *US Air Force Cost and Planning Factors, FY2008 Standard Composite Rates by Grade*, Air Force Instruction 65-503, Annex 19-2, Washington, D.C.: Secretary of the Air Force for Financial Matters (SAF/FMBOP), April 2007a.

————, Discoverer Data System, 2008a, not available to the general public. As of May 14, 2009:
https://tie65u01.okc.disa.mil

————, Manpower and Execution System, MPES, 2008b.

————, Deputy Chief of Staff for Installations and Logistics Directorate of Transformation (AF/A4I), *Logistics Enterprise Architecture (LogEa) Concept of Operations*, May 24, 2007b.

U.S. Department of Defense, *Quadrennial Defense Review Report*, Washington, D.C., September 30, 2001.

————, *BRAC Commission Action Brief*, Washington, D.C., September 1, 2005a.

————, *Air Force Transformation Flight Plan*, PBD 720, Washington, D.C., December 2005b.

U.S. Department of Defense and the Joint Chiefs of Staff, *Mobility Capabilities Study*, Washington, D.C., September 19, 2005, not available to the general public.

U.S. General Services Administration, Federal Supply Service, "Awarded Contractors Under Domestic Delivery Services," Web page, October 26, 2006. As of May 28, 2009:
http://apps.fas.gsa.gov/services/express/awarded-cont.cfm

U.S. House of Representatives Committee on Armed Services: Panel on Roles and Missions, *Initial Perspectives*, January 2008. As of May 12, 2009:
http://militarytimes.com/static/projects/pages/hasc_roles_missions0308.pdf